Carlos Cordova is Professor of Geography at Oklahoma State University and Visiting Scholar in Archaeology at Kazan Federal University, Russia. He obtained his PhD at the University of Texas at Austin under the supervision of Karl Butzer. He has undertaken geoarchaeological research in North America, Mexico, the Middle East, Southern Africa, and Crimea and the Black Sea Region, and is the author of *Crimea and the Black Sea: An Environmental History* (I.B.Tauris, 2016) and *Millennial Landscape Change in Jordan: Geoarchaeology and Cultural Ecology* (2007).

ENVIRONMENTAL HISTORY AND GLOBAL CHANGE SERIES

This important new series provides a much needed forum for understanding just how and why our environment changes. It shows how environmental history – with its unique blend of geography, history, archaeology, landscape, environment and science – is helping to make informed decisions on pressing environmental concerns and providing crucial insights into the mechanisms that influence environmental change today. The focus of the series will be on contemporary problems but will also include work that addresses major techniques, key periods and important regions. At a time when the scale and importance of environmental change has led to a widespread feeling that we have entered a period of crisis, the Environmental History and Global Change Series provides a timely, informed and important contribution to a key global issue.

1. *Documentary Records of Climate Change*, Astrid Ogilvie
2. *A Dictionary of Environmental History*, Ian Whyte
3. *The Mediterranean World: An Environmental History*, Neil Roberts
4. *Seeds of Empire: The Environmental Transformation of New Zealand*, Tom Brooking and Eric Pawson
5. *Cities: An Environmental History*, Ian Douglas
6. *Japan: An Environmental History*, Conrad Totman
7. *Crimea and the Black Sea: An Environmental History*, Carlos Cordova
8. *Geoarchaeology: The Human-Environmental Approach*, Carlos Cordova

Carlos Cordova

Geoarchaeology

The Human-Environmental Approach

I.B.TAURIS
LONDON • NEW YORK • OXFORD • NEW DELHI • SYDNEY

Dedicated to the memory of Professor Karl W. Butzer,
a source of inspiration for this book.

I.B. TAURIS
Bloomsbury Publishing Plc
50 Bedford Square, London, WC1B 3DP, UK
1385 Broadway, New York, NY 10018, USA

BLOOMSBURY, I.B. TAURIS and the I.B. Tauris logo
are trademarks of Bloomsbury Publishing Plc

First published in Great Britain 2018
This paperback edition published 2020

A catalogue record for this book is available from the British Library.

A catalog record for this book is available from the Library of Congress.

ISBN: HB: 978-1-7883-1301-8
PB: 978-0-7556-0677-1
ePDF: 978-1-8386-0860-6
eBook: 978-1-8386-0859-0

Series: Environmental History and Global Change, volume 8

Typeset by OKS Prepress Services, Chennai, India

Contents

Figures and Tables

FIGURES

TABLES

Acknowledgements

I am greatly thankful to Lisa Maher, Christopher Ames, Thomas Cox, Luc Vrydaghs, Yannick Devos, Christopher Morehart, Charles Frederick, Nicholas Dunning, Timothy Beach and Sheryl Luzzadder-Beach, who provided information to illustrate the topics of this book. I would like to express my gratitude to Gary Huckleberry, Aleksander Borejsza, Lisa Maher, April Nowell, Christopher Ames, and Frances Griffin, who read parts of the manuscript and provided helpful comments and suggestions, and to Michael Larson for help with some illustrations and maps. Finally, my special thanks go to David Stonestreet for his work and patience with editing this manuscript.

Introduction

Geoarchaeology as a field is an essential scientific approach to studying the human-environmental relationships in the past and the present. From its original conception as a series of geoscience techniques applied to archaeological research, it has become more than a multidisciplinary approach bridging archaeology and the geosciences; geoarchaeology has evolved to tackle problems related to society and environment of interest not only to archaeology but also to other fields. In part, this focus on the environment developed from the foundations of geoarchaeology established by Karl W. Butzer in his *Archaeology as Human Ecology* (1982), a book that to this day is widely cited in research papers and read in archaeological courses.

In his book, Butzer laid the foundations of the ecological context in archaeology, a concept that has been built on by a number of geoarchaeological studies in subsequent decades, during which new models of interpretation have emerged. These new models are in part a response to the technological advances and capabilities developed to tackle certain practical research problems and in part due to more recent environmental concerns related to rapid climate change and the current environmental crisis, particularly the rise of the Anthropocene paradigm, which has created debates that inevitably involve the idea of past and present societies and their ecological contexts.

The present book, *Geoarchaeology: A Human Environmental Approach*, brings together Butzer's ideas about the ecological context, named here "the human-environmental approach in geoarchaeology", and places them within the framework of current trends in geoarchaeological thought, then showing the different approaches to interpreting the record spanning from the earliest hominin environments to those of the contemporary past.

This book is intended to be a source of ideas for further discussion, not only among students but also by the geoarchaeological community in general. It is also addressed to academics and students in areas close to geoarchaeology: anthropology, geography, historical ecology and environmental history. It is not meant to be a geoarchaeology textbook or to replace the existing ones but, rather, to complement them.

For the most part, treatises and textbooks on geoarchaeology follow a geomorphological-soil model, structured in terms of environments – usually soils and sediments – and then the different settings (fluvial, eolian, coastal, cave deposits, etc.) (e.g., Waters 1992; Stein and Farrand 2001; Holliday 2004; Rapp and Hill 2006; Goldberg and MacPhail 2006); or an archaeological-geology model, structured in terms of technical aspects such as soil micromorphology, sediments, geophysics methods, sourcing and site taphonomy (e.g., Courty, Golberg and MacPhail 1992; Herz and French 1989; French 2003; Garrison 2016).

The present book, instead addresses archaeological topics from the human-environmental view, that is to say it focuses on the ecological context. For this purpose, the book rests on a cultural ecological base from where particular problems related to sites, settings, landscapes and environments are reviewed. For example, the chapter on the geoarchaeology of hunter-gatherer societies goes from the cultural concept to the different settings (i.e, fluvial, eolian, cave deposits, etc.). By so doing, this structure encourages non-anthropological practitioners to consider the cultural ecological aspects for their specialties, and helps encourage anthropological/historical-archaeological practitioners to have an idea of the connection between cultural aspects and the physical (natural) aspects involved in the study of geoarchaeological problems.

In seeking to present the human-environmental approach in geoarchaeology, this book begins with an epistemological analysis of the overall field of geoarchaeology, including its modern status, its theoretical and methodological foundations, and the definition of the *geoarchaeological record,* as opposed to the geological and archaeological record, including ways that geoarchaeologists interpret it. This epistemological introduction (Chapters 1–4) serves as a context for the main problems regarding the study of societies and their environment in space and time (Chapters 5–17), beginning with the contributions of geoarchaeology to human evolution and ending with the contemporary past.

During the past four decades, geoarchaeology has experienced a dramatic increase in practitioners, academic programs, and publications. In that process, geoarchaeology has developed a number of innovative methods and technological capabilities that have transformed the ways archaeologists reconstruct the past and the way geoscientists study the recent past and the present. But despite these developments little has been written about geoarchaeology itself. Most treatises and textbooks focus on technical matters, which focus is very important for training and diffusion, but they have practically nothing on theoretical and methodological aspects or on the very nature of the discipline, particularly in terms of its internal diversity, and the prospects for the future in the context of a rapidly changing world. Although in the 1980s and 1990s several articles addressed some of the epistemological and practical issues of geoarchaeology, practically nothing on the subject has been published lately, particularly during the present century.

Given this situation, this book positions the human-environmental approach in geoarchaeology in the context of the current nature of geoarchaeology, its practice, its practitioners, its theory, and its method. Chapter 1 addresses the question: Where is geoarchaeology now? Chapter 2 tackles an aspect of the field rarely addressed in the recent literature: theory and method. Chapter 3 addresses another aspect that has been taken for granted, yet which is important in terms of understanding the scope and the frontiers of geoarchaeological research: What is the geoarchaeological record? In response to this question, the author identifies not only the visible and invisible parts of the record but also their scalability in the different levels of abstraction used in geoarchaeological research: site, setting, landscape, and environment. Chapter 4 addresses issues in the interpretation of the geoarchaeological record, that is to say: how do we interpret the geoarchaeological record?

The human-environmental approach is discussed in Chapters 5–17. Chapter 5 discusses the nature of the human-environmental approach in geoarchaeology – the main theme of the book – and Chapters 6–17 address geoarchaeological issues from the human ecological point of view, and, more explicitly, cultural ecological perspective, stretching from research in early hominin sites, through cultural development stages such as hunter-gatherers, early agricultural societies, and complex societies, to more recent historical and contemporaneous societies. To link these topics with the aspects previously discussed in relation to the geoarchaeological record, I seek to address problems of interpretation in the topic chapters through the use of the fundamental levels of abstraction in geoarchaeology, that is, site, geomorphic/depositional setting, landscape, and environment

The rationale of presenting the topics of Chapters 6–17 in a structure of cultural ecological approach stems from two objectives. The first one is to distinguish the ways of interpretation across the spectrum of archaeological and historical contexts. The second objective is to provide those students and practitioners of geoarchaeology with academic backgrounds in the geosciences with a more cultural view of the problems they themselves try to solve.

Although the topics of Chapters 6–17 may sound more of a linear development, they are perhaps the best way to locate certain problems that non-archaeology and non-anthropology practitioners of geoarchaeology may relate to. In some cases, the topics do not fit in one of the chapters. For example, the geoarchaeological implications of nomadism do not fit into one chapter alone but several (Chapters 8, 9, and 11). The case of environmental crises, although assigned to Chapter 14, has important developments in other chapters. Complex societies are yet another case of a single chapter, but with implications and further discussions in other chapters.

To illustrate further the topics of Chapters 6–17, I have included in each chapter two practical case studies that are intended to be representative of the issues discussed in each chapter. One problem that may draw criticism is that about a third of the cases come from my own research and experience. In defence, I believe that one can provide more detail on a topic in which one has been deeply immersed. After all, most authors use their own research in their treatises to make a point or to convey complex ideas.

In addition to addressing relevant issues developed in the human-environmental tradition in geoarchaeology, this book attempts to bring to the forefront a series of issues that has been neglected or received little attention. These would be referred to by Karl Butzer (personal communication) as "those research issues that have fallen through the cracks." These issues include a variety of aspects such as ethnogeoarchaeology, phenomenological approaches in geoarchaeology, the use and misuse of modern analogs, and the role of geoarchaeology in pre- and proto-anthropic environments, and the geoarchaeology of the contemporary past, among others, as well as issues that geoarchaeologists are now taking on – such as the cultural aspects hidden in paleosols, experimental geoarchaeology, the application of computer simulation models – as well as emerging issues and approaches in the discipline.

One aspect that I feel is often taken for granted is the concept of "geoarchaeological record" as opposed to archaeological record and geological record (Chapter 3). In defining and conceptualizing the geoarchaeological record, I provide some aspects of interpretation that are rarely discussed in the literature (Chapter 4). These include aspects of causality, equifinality, time-transgressive phenomena, legacies, and visibility, invisibility, and absences in the record, among others.

Finally, this book is addressed to a variety of readers, from those actually practicing geoarchaeology – inside and outside academia – to those who are interested in environmental history or even disciplines allied with anthropology and sociology. In particular, those geoarchaeology practitioners with backgrounds in earth sciences and geography will find it a useful guide to interpreting aspects of cultural and human influence in the environment. It is also expected that the book will be used in geoarchaeology or environmental archaeology courses as a companion to traditional textbooks in geoarchaeology.

1

The Nature of Geoarchaeology

THE NATURE OF GEOARCHAEOLOGY AND ITS PRACTITIONERS

Where is geoarchaeology now?

It has been more than four decades since the term "geoarchaeology" was first used in print (Butzer 1973) and the first book bearing the word "geoarchaeology" in its title was published (Davidson and Shackley 1976). Since that time, geoarchaeology has experienced many changes, not only in the number of practitioners and publications but also in the diversification of specializations and the broadening of its scope. But rather than recount the history of geoarchaeology, which many authors do (e.g., Garrison 2016; Hill 2016; French 2003; Goldberg and MacPhail 2006; Rapp, Jr. and Hill 2006; Waters 1995), I believe it is more pertinent to discuss its most recent developments, its current status, and its prospects for the future, all of which are an important background to the main topic of this book: the human-environmental perspective in geoarchaeology.

In recent years, geoarchaeological research has begun to transcend topics of uniquely archaeological interest, as it has begun to touch on topics of societal and environmental issues (e.g., Beach et al. 2008; Butzer 2008; 2011; 2015; Benedetti, Beach and Cordova 2011; Wilson 2011; Cannell 2012). Furthermore, fields such as forensic geoarchaeology, historical geoarchaeology, and other areas related to the geoarchaeology of the contemporary past (Chapter 17) are matters of great relevance to high officials in institutions and the public. These new frontiers in geoarchaeology could easily serve as material to advance the discipline outside its academic circle.

Despite the advances in applications to archaeological, and in general to historical environmental problems, geoarchaeology still lacks a synthetic view of the discipline itself amid the number of specializations and approaches, a matter pointed out by several authors (Leach 1992; Killick and Goldberg 2009; Brown, Bassel, and Butzer 2011; Fouache 2013; Canti and Huisman 2015). This means that little has been done to unify the many concepts (including the definition of geoarchaeology itself), objectives and goals, while reconciling the different approaches developed by practitioners trained in different disciplines (henceforth referred to as the background disciplines). As a result, practitioners of geoarchaeology come each from their different disciplinary backgrounds, bringing on the one hand, their theoretical, methodological, and technical contributions, and on the other, their perspective.

The diversity of scientific backgrounds among practitioners is a topic very rarely addressed in the literature, despite its importance in the development and the nature of geoarchaeology. Therefore, this is perhaps one of the most important aspects to start with in terms of assessing the current situation of geoarchaeology. Other aspects that should follow are the practice itself and aspects of teaching and training, as well as the fast-developing subfields within the discipline.

The multidisciplinary nature of geoarchaeology

The creation of geoarchaeology as a research field was meant to bridge the gap between the human-focused field of archaeology and the empirical, nature-focused geosciences (Gladfelter 1981; Butzer 1982), a development that in turn resulted from the science-oriented philosophical movement of the New Archaeology (Rapp, Jr. and Hill 2006). Under this paradigm, adaptation of the methods of geology and other geosciences was welcomed in archaeological research as a way of solving a number of problems of interpretation of the archaeological record, particularly the formation processes (Butzer 1982). This led to the idea of collaboration between geoscientists and archaeologists, and eventually the inclusion of the geoscientist as a member of an interdisciplinary archaeological team. But because the geoscientist applies his/her knowledge and skills to archaeological research, the geoscientist is called a "geoarchaeologist."

One might assume that a geoarchaeologist is a specialist who, although trained in the geosciences, is knowledgeable of archaeological methods and research problems. Likewise, it is assumed that archaeologists understand the work done by geoscientists and other specialists. But in practice this is not always the case. Often a geoarchaeologist whose specialty is soils and geomorphology is asked to solve a question related to lithic and clay sourcing. In this case the problem is that the failure to recognize the specialization leads to overrating – or in some cases underrating – the role of the specialist in research. Misunderstandings on the side of geoscientists are apparent in many situations where interpretations are done without regard of the nature of human scales.

To address these differences of perspective involves a mutual understanding of the methods, procedures, and even theoretical aspects across the disciplines involved in archaeological research. With regard to this issue two fundamental questions may be raised:

(1) How much archaeology and anthropology should a practitioner of geoarchaeology with a geoscience background know?
(2) How much geoscience should a practitioner of geoarchaeology trained in archaeology and anthropology know?

The issue related to the questions above has been discussed in the literature, with calls on one side being made to geoscientists working in the archaeological research to grasp the cultural aspects inherent to archaeology and anthropology (Ellery 2004; Frahm 2004; Leach 1992; O'Sullivan 2008); whilst, on the other side, calls have been made to archaeologists to understand aspects of earth sciences that concern archaeological interpretations (Butzer 1982; Cordova 2007; Fouache 2013; Huckleberry 2000; Walsh 2004;). This is perhaps a matter that not only requires more discussion among the members of the geoarchaeological community but is also a curricular matter that should be considered in college and graduate programs that offer geoarchaeology as a specialization. Specialization sometimes comes with the academic background, and currently has created such a diversity in geoarchaeology that one is sometimes left with the idea that there is not one but many geoarchaeologies. Archaeological geology, environmental archaeology, archaeometry, Quaternary ecology, and many other fields are sometimes other ways of naming geoarchaeology or fields that overlap geoarchaeology (Canti 2001; Fouache 2013).

This diversity of specialized fields and approaches needs to be sorted out and defined under the umbrella of a synthetic approach to geoarchaeology that recognizes specialization but at the same time recognizes the role of geoarchaeology in relation to broad methodological problems and applications. Recognizing the diversity in geoarchaeology is thus the first step to understanding the field as a whole.

Archaeological geology or geoarchaeology?

Archaeological geology is the term that defines the field in which geology and archaeology collaborated before the term geoarchaeology was coined in the mid-1970s (Rapp, Jr. 2007; Garrrison 2016). In its original conception, archaeological geology meant the application of methods of geology to archeological problems (Butzer 1982; Rapp, Jr. and Gifford 1982). Although archaeological geology is sometimes used interchangeably with geoarchaeology, the latter is a much more inclusive terminology, which probably explains its more frequent usage today (Canti 2001; Garrison 2016). However, geoarchaeology not only includes geological fields but also geographical and biological ones (Hill 2016), which means that the term archaeological geology should be reserved only for the direct use of geological methods in geoarchaeology.

Despite the preference for geoarchaeology, the term archaeological geology has not fallen out of use. It is still used in titles of volumes (e.g., Herz and Garrison 1989; Garrison 2016) and as a broad field of research in geology departments and organizations, as is the case with the Archaeological Division of the Geological Society of America (Rapp, Jr. 2011). This society held its first session on archaeological geology in 1973, and the division was formally created in 1977 (Rapp, Jr. 2007; Hill 2016). In those years, the term geoarchaeology was still not as popular as it is today. Interestingly, however, most members of the Archaeological Geology Division identify themselves as geoarchaeologists even though many are not geologists. However, in 2017 the division finally changed its name to the Geoarchaeology Division.

Although less used in the literature, the term archaeological geology has not disappeared. Instead, it designates an approach or tradition within geoarchaeology. This tradition represents mainly the application of geological methods to archaeological research, as opposed to other fields of research in geoarchaeology which focus on other research subjects – a matter that is described in more detail later in this chapter.

Environmental archaeology and geoarchaeology

Environmental archaeology is a multidisciplinary field created within archaeology, more common in the English-speaking archaeological world than anywhere else. Geoarchaeology is generally accepted as one of the branches of environmental archaeology along with archaeozoology, archaeobotany, and Quaternary paleoecology (Wilkinson and Stevens

2001). But despite the acceptance of geoarchaeology in the general framework of environmental archaeology, many geoarchaeologists do not consider themselves environmental archaeologists.

It is important to emphasize that the conceptualization of environmental archaeology is different on both sides of the Atlantic. In North America, environmental archaeology is more closely seen as the study of past human ecosystems – an idea that is more akin to ecology and less to other fields (O'Connor 1998; Dincauze 2001). In Britain, where the term is more popular, environmental archaeology is a field dominated by biologists, and its basis is more in biogeography than in ecology (see Chapter 5 for the influences of these two fields in archaeology). Interestingly, it is in Britain where most geoarchaeological work is seen as environmental archaeology (Canti 2001; Wilkinson and Stevens 2001; Reitz and Shackley 2002). In North America, however, the relationship between environmental archaeology and geoarchaeology is weak in theoretical terms but relatively strong in practice, particularly among geoarchaeologists adhering to the human-environmental tradition (see next section).

THE THREE MAJOR TRADITIONS IN GEOARCHAEOLOGY

This diversity of academic backgrounds and technical specializations has produced a variety of ways of approaching archaeological problems, with some focusing more on the geochemical aspect, others more on the analysis of paleosurfaces, and others more on the ecological aspects (Benedetti, Beach and Cordova 2011). Because each of these approaches has inherited some theoretical and methodological aspects from traditional disciplines such as geology, geomorphology, and geography, they can also be referred to as traditions within geoarchaeology.

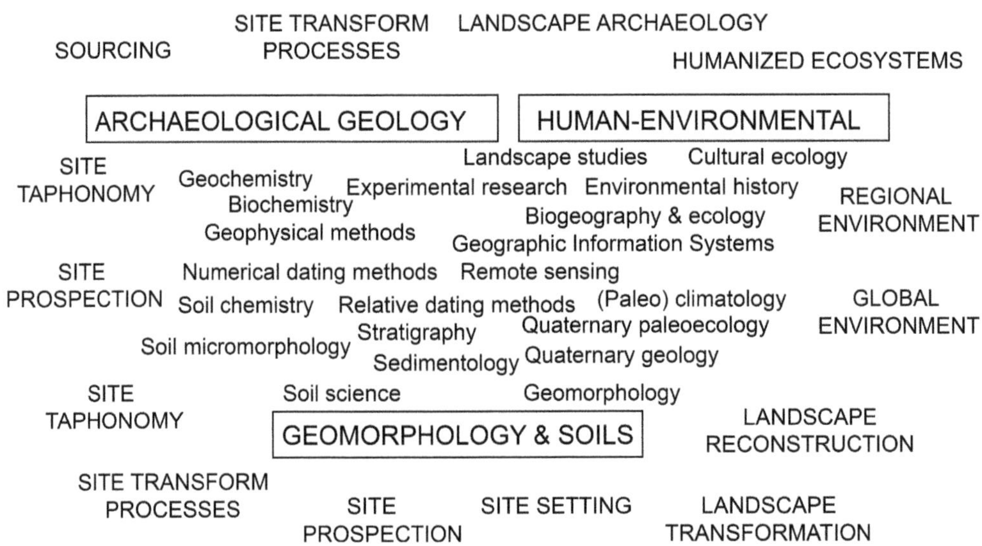

Figure 1.1. The three approaches or traditions in geoarchaeology in the context of technical fields (center) and research focus (outer area).

The three geoarchaeological approaches or traditions may be recognized as:

- archaeological geology (the geochemical aspect),
- geomorphology and soils (the paleosurface aspect),
- and the human-environmental approach (ecological and cultural aspects).

Each differs in their background disciplines and in the specializations among their practitioners (Figure 1.1).

The division into three approaches or traditions, however, should not be seen as suggesting separate approaches, where one can place research and researchers; but rather as research focuses, with no defined boundaries, as some practitioners may be conversant in fields with more than one tradition. The three traditions do not constitute schools of thought but, rather, broad fields of specialization with particular approaches to different research problems within geoarchaeology.

The archaeological geology tradition

Archaeological geology has contributed directly to archaeology from the broad field of geology, a tradition that predates the coining of the term geoarchaeology (Rapp, Jr. and Gifford 1982). Contributions to this field are notable in several treatises, among which the most important are Shackley (1975), Herz and Garrison (1989), Stein and Farrand (2001), French (2003), Rap Jr. and Hill (2006), and Garrison (2016), among others. Furthermore, aspects of soil micromorphology that in part are archaeological geology and in part soil science include the treatises by Courty, Goldberg, and MacPhail (1989) and Goldberg and MacPhail (2006). One may include in archaeological geology the applications of sedimentary environments to geoarchaeology, a topic which is represented in some chapters of compilations by Goldberg and Holliday (2001) and Stein and Farrand (2001). Aspects of archaeological geology are not confined to geochemical aspects but include a broad range of fields such as absolute dating techniques, geophysical prospection, stratigraphy, and biogeochemistry (Figure 1.1). For the most part, its practitioners are trained in geology and geophysics, although those with a background in archaeology are not excluded.

The geomorphology and soils tradition

The geomorphology and soils tradition, hereinafter the geomorphology-soils tradition, focuses primarily on the evolution of the landscape and the aspects of sedimentology and pedogenic development that affect the formation and transformation of archaeological sites (Gladfelter 1977, 1981; Hassan 1979; Stein 2001a). For the most part the view of soils is different from that of soil micromorphology in the sense that the most important role of soils is linked to geomorphic stability (Mandel 2000; 2008; Holliday 2004).

This approach has strong links with archaeological geology, particularly in aspects of stratigraphy, which are not only a link to sediments but also to soil complexes (pedocomplexes). Empirical studies in modern environments are important in this area, as reflected in many treatises, edited volumes, and monographs (e.g., Waters 1992;

Brown 1997; Holliday 1997; Pollard 1999; Mandel 2000; Goldberg and Holliday 2001; Wilson 2011). Examples of this kind include the study of depositional environments through facies analyses and facies models, a method directly taken from geology (Feibel 2013), but with a strong link to geomorphological dynamics.

The academic background of most of the practitioners in this tradition is for the most part in geology and geography, given the bridging that geomorphology creates between these two disciplines. But practitioners with backgrounds in soil science, and to a lesser degree in archaeology and anthropology, use the geomorphology-soils approach in their geoarchaeological research. Interestingly, of the three geoarchaeological traditions, the geomorphology-soils tradition has been the most widespread in North America (Waters 1995; Benedetti, Beach and Cordova 2011).

The term *archaeogeomorphology*, which has been used sparsely in the geomorphological literature, does not necessarily refer to the approach referred to here as the geomorphology-soils tradition. The term was first proposed in 1992 (Wandsnider 1992), and two decades later was reviewed, reformulated and proposed as a field of geomorphology (Thornbush 2012). Although the term has, in some instances, been used as synonymous with geoarchaeology, it has also been used to refer to geomorphology applied to archaeological problems (Thornbush 2012; Nicu 2016). But in neither case was the term successful in being used in the archaeological and geoarchaeological literature, and only very sparsely in geomorphology and applied physical geography papers (Thornbush 2012; Hesse 2014). Use of the term has varied and sometimes it has been used for specific cases in specific fields – as, for example, the reconstruction of relief in archaeological studies of battlefields (Hesse 2014).

If archaeogeomorphology is used as synonymous with geoarchaeology, the scope of the term fails to recognize the role of other geosciences in the broad scope of geoarchaeological research. However, if it is used to refer to the geomorphology and soils approach described here, the term would need to absorb the soil and stratigraphic work that accompanies geomorphological work in geoarchaeology. Certainly, the term would first have to be popularized among the diverse pool of practitioners of geoarchaeology.

The human-environmental tradition

The human-environmental tradition has its origin in some of the early stages of geoarchaeology as several authors began to incorporate ideas from the broad ecological paradigm into the interpretation of the geoarchaeological record (e.g., Butzer 1975, 1980a; Hassan 1979), reaching its consolidation in 1982 with the publication of *Archaeology as Human Ecology* by Karl W. Butzer. This tradition could be referred to as the "ecological contextual tradition," but for the purpose of a short and broad-meaning name it is here called the "human-environmental tradition."

Despite maintaining a strong base in the geosciences, the human-environmental tradition is more strongly tied to human ecology and cultural ecology. It also maintains a strong connection with the other three areas of environmental archaeology: archaeobotany, archaeozoology, and paleoecology. Although geographers, archaeologists, and anthropologists make up the bulk of practitioners associated with the ecological and contextual approach, other specialists are often involved. Furthermore, it is not uncommon for most practitioners to have a foot in the geomorphology-soils tradition whilst also researching aspects of human ecology. The human-environmental tradition is the main focus of this book, and many aspects of it are discussed in Chapters 5–17.

The meaning of the three geoarchaeological traditions

The academic background and the specialization in a field of research often define the tradition (or approach) a practitioner follows in geoarchaeology (Figure 1.1). But despite the different approaches, there are common aspects that practitioners try to explain, particularly in relation to formation processes and human behavior, both of which are archaeological issues. The difference among the three approaches is how questions and problems in archaeological research are approached by geoarchaeology (outer area of the diagram in Figure 1.1). In practical terms, the division into three approaches are, on the one hand, a division of labor within geoarchaeology.

At the same time, the traditions constitute broad areas that are characterized by the specialized scientific fields in geoarchaeology (center of the diagram in Figure 1.1). Although practitioners may be openly associated with one of the three traditions, it is possible to be part of two or all three, or perhaps claim no association with any of them. For example, soil micromorphology, although strongly associated with archaeological geology, it may play an important role in the study of geomorphic processes.

THE FIELD AND ITS STATUS IN THE SCIENTIFIC REALM

Is geoarchaeology a scientific discipline?

The goals and scope of geoarchaeology were well defined during the late 1970s and early 1980s and have been amply discussed in numerous papers in recent decades (e.g., Leach 1992; Rapp, Jr. and Hill, 2006; Huckleberry 2000; Canti 2001; Butzer 2008; Wilson 2011; Canti and Huisman 2015; Garrison 2016) and in articles in the recently published *Encyclopedia of Geoarchaeology* (Gilbert 2016). But despite the clear goals, there is still the problem of defining the status of geoarchaeology in the realm of the sciences, that is to say, whether it is a sub-discipline or field within archaeology or geology, or a bridge field between archaeology, geology, geography, and geomorphology.

The original conception of geoarchaeology in the early 1970s was as that of a technical field that linked the capabilities of geology to solving archaeological problems (Renfrew 1976). But soon geoarchaeology began to be seen not just as a technical field, but as a scientific enterprise (Gladfelter 1977; Hassan 1979; Butzer 1980a; 1982; Rapp, Jr. and Gifford 1982), and by the time *Geoarchaeology: An International Journal* was founded in 1986, geoarchaeology had reached a broader academic field with a more defined purpose (Leach 1992; Waters 1992).

In the first decade of the new century, geoarchaeology became more recognized in geoscience and archaeological organizations including the International Association of Geomorphology (IAG), the International Quaternary Association (INQUA), the World

Archaeology Congress, and the International Geological Congress, as well as in other national organizations such as the Society for American Archaeology (with the creation of the Geoarchaeology Specialty Group) and the Association of American Geographers (with the Paleoenvironments Specialty Group).

The first decade of the twenty-first century saw the publication of new treatises and a multiplication of papers in *Geoarchaeology: An International Journal*, as well as in the *Journal of Archaeological Science, Geomorphology, Quaternary International, Quaternary Research, Quaternary Science Reviews*, and a number of archaeological and geoscience-related journals. Additionally, sessions in meetings and independent meetings on geoarchaeology increased considerably, bringing more and more students and academic and non-academic practitioners to share results and discuss a variety of aspects of the geoarchaeological discourse. In general, in the second decade of this century, geoarchaeology has developed a dynamic structure (Fouache 2013; Gilbert et al. 2016) and more involvement in research including aspects of modern relevance such as the anthropocene, environmental change, and climate change.

But despite its dynamic development and the clear goals and scope, there is still the question of its status among the disciplines, that is to say, the question as to whether geoarchaeology is truly a discipline, a field in archaeology or geology, or just a research approach. Although no paper has fully discussed this issue, the perceived scientific status can be deduced from the definitions of geoarchaeology, which in turn present different points of view. Some definitions see geoarchaeology as an interdisciplinary research approach (e.g. Butzer 1982; French 2003; Rapp, Jr. and Hill 2006; Butzer 2008), others as a discipline (e.g., Goldberg and MacPhail 2006; Wilson 2011; Garrison 2016; Maher 2017), others as a sub-discipline (Stein 2001b; Gilbert et al. 2016) and sub-field (Butzer 2008; Fouache 2013), and others as a science (Huckleberry 2000; Engel and Brückner 2014; Hill 2016). When it is mentioned as an interdisciplinary field, it is often associated with the geosciences and archaeology; but when it is a sub-discipline or sub-field, the implication is that it belongs to archaeology (Butzer 2008); but if it is mentioned as a discrete scientific discipline, the implication is that it is not a subfield of any major discipline.

In the strict semiotic sense of the word "science" (sensu Mario Bunge 2002), it is acceptable to consider geoarchaeology a science because it uses the scientific method. But in the broader sense of the word science, as a scientific discipline, geoarchaeology lacks independence, visibility, and acceptance that most major sciences (e.g., geology and geography) and their scientific disciplines (e.g., geomorphology, paleontology, biogeography and climatology) have. This explains why most definitions see it as a sub-discipline or simply a research approach. This situation, however, could change, as it happened to other scientific disciplines (e.g., geomorphology), which now have more acceptance and visibility in the broader scientific realm.

Geomorphology for the sake of comparison

One way to evaluate the scientific status of geoarchaeology is to compare it with geomorphology, a field that originated in part as an approach and sub-field of geology and geography, but today enjoys more independence and recognition in the scientific realm. Although developments of geoarchaeology and geomorphology occurred within different scientific and social contexts, it is worth referring to the case of geomorphology because it is also a discipline that faced problems of recognition and divergence among its practitioners, but also some sort of convergence on synthetic aspects (Pitty 1982; Walker and Grabau 1993; Rhoads and Thorn 2011).

Geomorphology was born in the early twentieth century as a bridge between geology and geography. In geology it became the way to link structures and rocks to the evolution of the landscape, and in geography it became the topographic basis for explaining other phenomena (Butzer 1976). Geology provided the lithological base, while geography, as seen in the development of William Morris Davis's geographic cycle, provides the relation with the modern landscape (Cordova 2016b).

But towards the middle of the century the discipline faced an identity crisis, in part due to the conservative view that strongly aligned geological time scales (Butzer 1980b). But neither the time scales served geology well, nor explained many processes at the scales common to geography. The latter meant also a disconnect with other aspects of the spheres, mainly the biosphere. This also meant a stagnation in methodological terms, which failed to explain aspects of landform and landscape change in more credible models (Pitty 1980). There were also problems that impeded geomorphology to be more widely recognized.

In *L'épistémologie de la géomorphologie*, Alain Reynaud (1971) explains the origin and solutions problems of identity, as being part of either geology or geography. Interestingly it was during or shortly the publication of this book that geomorphology began to take off in popularity and academic support, much of it in the context of the positivistic ideas that stimulated quantitative research and modeling particularly by British and American geomorphologists and to a lesser extent French, German, and Russian (Pitty 1982; Rhoads and Thorn 2011). This new wave saw the ingrown development of theoretical and mathematical models to study the evolution of the landscape and methodological capabilities of the emerging technologies at the time (e.g., computer models and the rapidly evolving field of remote sensing). Through the 1970s and 1980s, geomorphology created a lexicon of its own, borrowing and adapting terms from systems theory such as threshold, steady state, and recovery, among others (Cordova 2016b).

Although many national societies existed since the middle of the century, the birth of the International Association of Geomorphologists in the late 1980s marked an important step in the recognition of this scientific field (Walker and Grabau 1993). Although geomorphology academic programs are still attached to either geology or geography, or in other cases to environmental sciences, it enjoys are more recognized status with a large number of "geomorphologists," most of which are affiliated with national and international organizations, and with recognition as a scientific field in many universities and research institutions around the world.

What this example tells us about the future of geoarchaeology is that one day the kind of revolution geomorphology saw in the early 1970s shortly after the time of Alain Reynaud's (1971) book, may occur or perhaps is occurring in geoarchaeology. Perhaps in one or two decades geoarchaeology will attain the degrees of independence and scientific maturity similar to geomorphology has today. It is also important to recognize that the scientific independence and maturity in geomorphology was created through the development homegrown theory and method (Rhoads and Thorn 2011). Geoarchaeology, a field straddling geosciences and archaeology, should perhaps look into developing robust homegrown theories and methods, which along with diffusion in and outside the sciences, should create possibilities of advancing the field into an academic discipline.

Geomorphology has gained its due recognition beyond the confines of their scientific circles, because geomorphologists have worked hard on making their field more visible through the diffusion of its scientific capabilities and usefulness to a broader series of problems of society, the environment, and education. In contrast to geoarchaeology, geomorphology has better chances of getting funded either by the Geosciences or the Geography and Spatial Science programs, or by multidisciplinary programs. Perhaps the geomorphological example should be taken seriously in geoarchaeology, particularly in North America, where it is far from being recognized as a significant field of research by government and state funding agencies.

Geoarchaeological theory, method, discourse, and lexicon

The rise of geoarchaeology in the past four decades is evident in the growth of practitioners, published research, specialized meetings, and focus groups in national and international geoscience and archaeological organizations. But of scientific independence and maturity, as discussed in the previous section, one question remains: is there a geoarchaeological theory and method?

A satisfactory answer to this question is complicated and lengthy, deserving some details and background information, which is provided to a certain degree of detail in Chapter 2. But a short answer can be stated as follows: not being a scientific discipline, but a multidisciplinary field or research or approach, geoarchaeology has borrowed methods from the parent and sponsoring disciplines such as archaeology, geology, geography, geomorphology, soil science, and to a lesser degree ecology. However, geoarchaeology has thus far created one the one hand a scope and goals, and on the other a robust discourse and a lexicon, all of which create the foundations for the development of homegrown theories and methods.

In philosophy of science, *discourse* means the language that scientists use, which includes codified terms (lexicon), specialized writing or means of communication, and debates (Von Engelhardt and Zimmermann 1982). In essence, the discourse includes the means of communications among scientists and their products, which often are seen in publications, debates, panels, meetings, maps, and in modern times, of course, blogs and materials in the social media. In this aspect, geoarchaeology has seen great advances. For this reason, the creation of *Geoarchaeology: An International Journal* was the first step in creating a focused and specialized forum of discussion. Additionally, creation of divisions within national and international organizations, and the number of meetings devoted to geoarchaeological problems, enhanced the geoarchaeological discourse.

With regard to the geoarchaeological lexicon, an ever-growing introduction of terms have been used for common understanding between geoscientists and archaeologists. This has come to solve the problem of usage of terms with different meanings in both sides (e.g., horizon, intrusion, structure, etc.), or meaning only in archaeology (e.g., sterile layer, primary and secondary context, floor, etc.) or only in the geosciences (e.g., parent material, calcrete, illuviation, diagenesis, etc.). The publication of the *Encyclopedia of Geoarchaeology* (Gilbert 2016) is a great step forward in defining and consolidating the geoarchaeological discourse and lexicon of the second decade of the twenty-first century.

The identity challenge

Despite its current popularity and acceptance among archaeologists and geoscientists, geoarchaeology faces some identity problems in the broader academic environment, government institutions, funding agencies, and the public. As a result, geoarchaeology fails to be given its due recognition in the academic structure of most universities and other research institutions, one problem that Karl W. Butzer addressed multiple times in talks, in meetings, and in some of his most recent publications (Butzer 2008, 2015). This situation calls for the need to advance geoarchaeology not only within the academic ranks of the disciplines involved (i.e., archaeology, anthropology, and the geosciences) but also within visible distance of university officials, policy makers, and the public in general.

The lack of broad academic recognition of geoarchaeology is reflected in the mechanisms for funding procurement, as is the case National Science Foundation (NSF) –the major source of federal funds for scientific research in the USA. The NSF is structured into several divisions, which in turn are divided into programs. Often archaeologists obtain funds from the Archaeology and Archaeometry Program, geologists from the Geoscience Program, and geographers from the Geography and Spatial Science Program. In the case of geoarchaeology, the majority of funding has come from the Archaeology and Archaeometry Program, but often as part of a broader archaeological research project. This makes perfect sense as long as geoarchaeology is part of the conglomerate of scientific fields in an archaeological project.

However, some geoarchaeological research projects on their own often find difficulty fitting the general norm of an anthropological-based archaeological model, despite their importance as scientific enterprises with potential applications to archaeological problems. Thus, their chances of even been considered for funding through the Archaeology and Archaeometry Program are small, if not non-existent. Similarly, the same prospective research projects find difficulties for funding through the Geosciences and the Geography and Spatial Science programs because they sound too archaeological.

In theory, however, geoarchaeology with its multidisciplinary dimension could qualify for some of the NSF interdisciplinary programs, but often such programs are either tailored to specific targets or involve divisions that are completely unaware of the scientific capabilities of geoarchaeology. Aspects of disproportion between public funding support for geoarchaeology in the USA and Europe have been pointed out (Killick and Goldberg 2009; Canti and Huisman 2015; Killick 2015) with the disproportion resulting in fewer technological and methodological contributions from the USA, despite the amount of work done there, with respect to Europe and other first-world countries (Canti and Huisman 2015). Although the problem seems to be the North American position of archaeological science in general in relation to anthropological/humanities-based, as opposed to the more science-oriented European counterpart (Killick and Goldberg 2009; Killick 2015), the problem seems to be that geoarchaeology is still for the most part invisible. It is also noticeable that other countries, particularly in Europe, funding agencies target topics, not disciplines, which makes easier to those multidisciplinary and non-aligned projects to qualify for funding.

Rather than decrying problems of recognition, perhaps it is better to explore the very nature of geoarchaeology, its internal problems, and their solutions. Thus, it is the purpose of this chapter and the next one to discuss various aspects of theoretical and methodological background that can be exploited towards building a stronger field. Perhaps geoarchaeology should follow the steps of geomorphology, which, as discussed above, was able to emerge from the shadows of geology and geography. But beyond advances in the scientific and academic arenas, it is through diffusion and outreach that geoarchaeology should find its way to a more visible discipline.

PRACTICE, TRAINING, AND THE RAPIDLY EVOLVING SUBFIELDS

Integration of geoarchaeological research in archaeology

The incorporation of geoarchaeological research into the main archaeological research project as a topic was extensively discussed in the years after the creation of the terms geoarchaeology and archaeological geology (e.g., Hassan 1979; Butzer 1982; Gladfelter 1977, 1982; Rapp, Jr. and Gifford 1982). In more recent years, however, matters regarding integration are still being discussed, but now under a broader theoretical and practical ways (See, for example, discussions in Leach 1992; Huckleberry 2000; Holliday 2004; Schuldenrein 2006. 2007; Wells 2001; Fouache 2013). It is clear that such discussions gravitate around the issue of research coordination and integration of data, which represents some challenges particularly because the way geosciences and archaeology traditionally work.

Normally, archaeological tasks, defined as the *archaeological process*, are divided into survey, excavation, laboratory, and publication (Peregrine 2001). Geoarchaeologists, who in most cases work in agreement with schedule and procedural needs of archaeology, follow the archaeological process. To that purpose, Karl W. Butzer (1982) proposed the following phases of geoarchaeological research: field research at the site and the landscape, laboratory, revision strategies, and multidisciplinary integration. The way geoarchaeological and archaeological tasks are coordinated varies, but in general they tend to be somehow independent, though still maintaining channels of communication and coordinated timing (Figure 1.2).

Variations to the phases proposed by Butzer (1982) include aspects of the nature of the work itself, as in CRM geoarchaeology (Schuldenrein 2007), or whether the work is more focused on site, landscape, or environment (Stafford 1995; Wells 2001; Fouache 2013), and the particular nature of research. For example, one can expect different coordination and overall procedures and schedules between, say, research focusing on Lower Paleolithic sites and that focusing on urban historical archaeology. Likewise, a study of an underwater site (e.g., a shipwreck) should require modifications to research tasks and phases. Specialized research in geoarchaeology, requiring perhaps work at different times or different localities, can also modify the coordinated tasks depicted in Figure 1.2, as is the case of sampling at a particular time of the year, or work in off-site research, or regional sampling.

The coordination between geoarchaeological and archaeological can sometimes have its problems and conflicts, particularly in terms of communication and time coordination, or problems related to time and budget constraints, which are common in multidisciplinary research projects. There are at least three issues inherent to the coordination between geoarchaeological and archaeological research. The first one is the lack of geoarchaeological objectives or goals in an archaeological project. For example, there are cases where a geoarchaeologist is placed in a research design because all archaeological projects have to

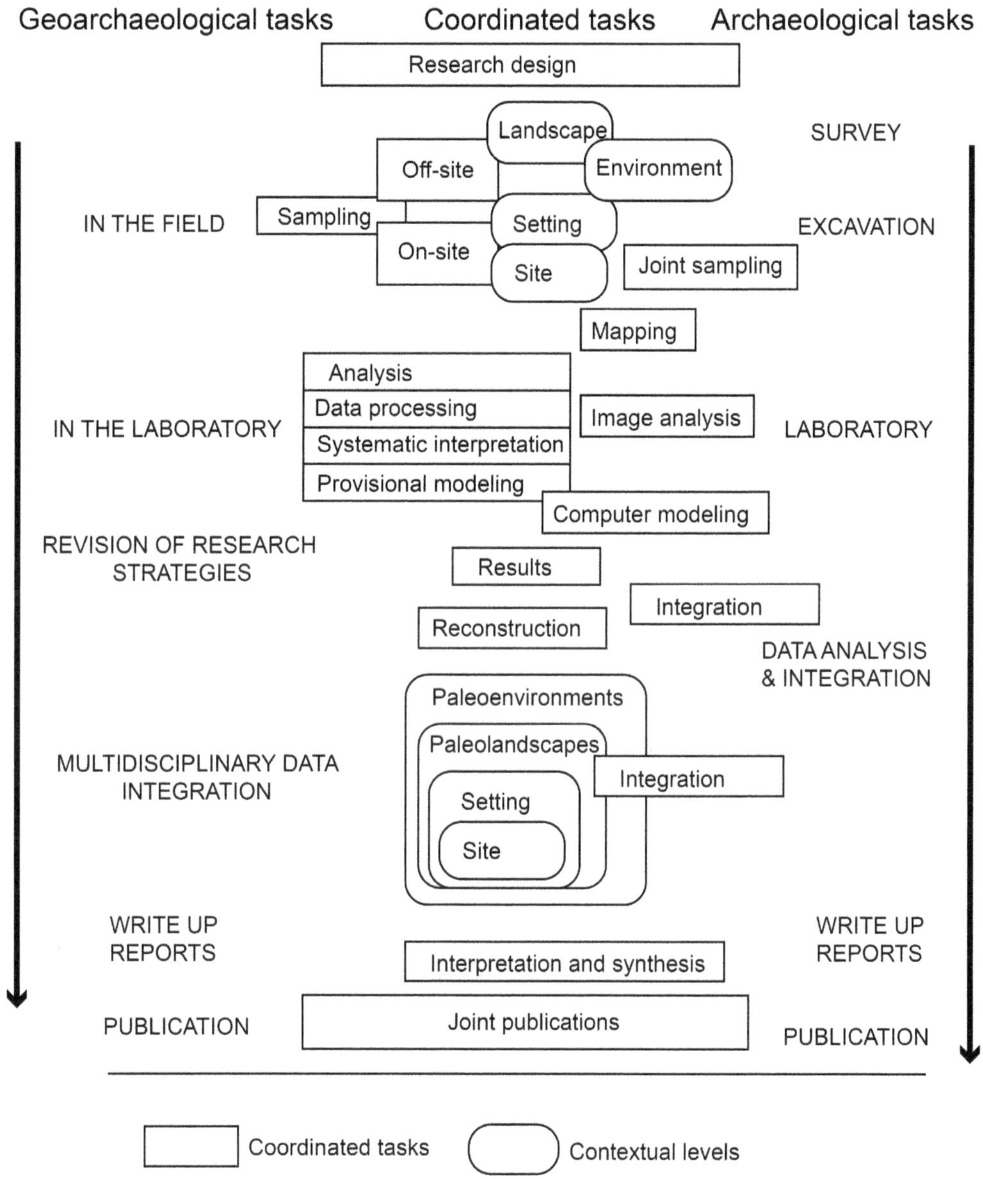

Figure 1.2. The methodological relation between the archaeological research process (right), and the geoarchaeological tasks according to Butzer (1982) (left), indicating contexts (site, setting, landscape, or environment) and coordinated tasks.

have one or because it will make the project look more scientific, but the goals of geoarchaeological research are not defined. This problem is becoming less common, since now most archaeologists understand the role of geoarchaeology, and the potential problems it can solve in a research project, particularly if particular objectives are established to test archaeological hypotheses.

The second issue is related to the overstatement of what a geoarchaeologist can technically do in a project. As discussed above, there are various fields of technical specialization in geoarchaeology (Figure 1.1), and a practitioner is often trained in only one, or maybe two, such fields. For example, a geoarchaeologist trained in soil geomorphology may not be of great help when the problem is to study chemical residues, lithic sourcing, or ceramic mineral composition. If objectives and goals are defined, then it will be clear what geoarchaeological problems can be tackled. Thus, it is not uncommon then that a project would have more than one practitioner of geoarchaeology.

The third issue deals more with the mutual understanding of archaeological and geoarchaeological research procedures. Each field has its *modus operandi* and particular steps of research to solve specific problems. For example, geoarchaeology can provide information about paleoenvironments, but that task cannot be done only at the site; it has to include off-site tasks (Stafford 1995; Wells 2001; Fouache 2013). Although off-site geoarchaeology is sometimes a source of conflict, particularly among more traditional archaeologists, it is a task that is may be essential particularly in terms of prospecting new archaeological findings and reconstructing paleolandscapes and paleoenvironments associated with the site (see Chapter 4). Off-site research, however, can be a formidable challenge in CRM archaeology, where the extent of work is tied to a contract.

It is important to accept that conflicts about methodologies in a team are possible, particularly between the anthropology-minded or history-minded archaeologist and the geoscience-minded geoarchaeologist (e.g., Leach 1992; French 2003; Holliday 2004; Fouache 2013). The solution to these problems lies in the proper dialog, clarification of goals, and proper planning at the research design stage (Figure 1.2).

Other potential problems in the coordination of geoarchaeological and archaeological tasks may arise toward the end, particularly in the integration of multidisciplinary data. This not necessarily refers to the integration of geoarchaeological and archaeological data, but also of other fields, particularly archaeobotany, archaeozoology, and archaeobotany. The key to a successful integration lies in what Dena Dincauze (2000) calls the "three-C goal," that is, complementarity, or integration of different data sets; consistency, or compatibility between reconstructions; and congruency, or mediation between data sources. On paper this looks easy, but in practice this can be problematic. Therefore, the mutual understanding of the other multidisciplinary fields is a must for all participating specialists. In the long term these problems can be avoided through courses, training, broad interdisciplinary curricula in academic programs, and multidisciplinary workshops, which may inform students of one discipline about the potentials of other disciplines in a multidisciplinary project.

Teaching and training in geoarchaeology

In the early days of the use of the term geoarchaeology, if not earlier, some authors proposed modifications of geoscience and archaeological programs to adapt the training of students in geoarchaeology (e.g., Butzer 1973, 1975; Hassan 1979; Gladfelter 1981). In the subsequent decades the literature began making a distinction between teaching geoarchaeology and training in geoarchaeology (e.g., Donahue and Adovasio 1985; Huckleberry 2000; Rapp, Jr. and Hill 2006). In this case, teaching involves offering geoarchaeology courses to train archaeologists in geoscience matters or geoscientists interested in geoarchaeology on archaeology and anthropology. Training, on the other hand, means preparing students in geoarchaeology as a career, which involves more than courses, but a curriculum often offered as academic programs.

In North America, a number of archaeology programs (undergraduate and graduate) that are part of broad anthropological programs still do not include a course in geoarchaeology, or even environmental archaeology, which means that many archaeologists are graduating with only a slight idea of the work of geoarchaeology and overall archaeological science (Killick 2015). In some cases, students take elective courses in geomorphology, geology, and soil science, but that does not mean they will bring together archaeological and geoscience problems, particularly those of scale. However, whether or not archaeology students follow the path of geoarchaeology, they should have a background in geoarchaeology, because it is now an essential part of archaeological research.

In other parts of the world, particularly in Britain, the teaching of geoarchaeology courses has been taken more seriously, although this is perhaps the result of more science-oriented programs. In other European countries, including those of the former Soviet bloc, where archaeology is more linked to history, often the role of geoarchaeology is seen as a field more in the geosciences, except perhaps where archaeology is linked to prehistory, in which case there is more interest on the part of archaeologists in getting involved in geoarchaeology. In Australia, where most archaeology is prehistoric, the geoarchaeological interest is more important and more developed into archaeology curricula.

But despite the slow development of geoarchaeology curricula, the end of the twentieth century already saw degrees being offered with specialization in geoarchaeology (Huckleberry 2000). In the USA they are normally granted by anthropology, geology/geoscience, and geography, or as part of multidisciplinary programs. The Archaeological Geology Division of the Geological Society of America has recently updated on their website the number of graduate programs that form geoarchaeologists, some of which are in anthropology, geology, and geography departments, with a few inter-disciplinary (or inter-departmental) programs. A similar trend has been seen in other countries, particularly in Britain, where a number of programs, mostly in archaeology, are offering courses and degrees with specializations in geoarchaeological fields.

Unfortunately, topics of teaching and training in geoarchaeology are rarely discussed in meetings, and even less in publications, despite its importance in the advancement and diffusion of geoarchaeology. It is then the responsibility of the geoarchaeological communities to work on facilitating a synthetic course on geoarchaeology in all archaeology programs and to establish a solid curriculum that prepares archaeology and geoscience students for geoarchaeological research.

Academic, non-academic, and CRM geoarchaeology

The practice of geoarchaeology is traditionally divided into academic, usually associated with universities, and the non-academic, usually associated with government institutions and private consulting companies. Most of the non-academic geoarchaeological work is done around CRM (Cultural Resource Management) work, which concerns the broad fields of salvage archaeology and environmental impact assessment.

In the USA, the practice of CRM archaeology grew considerably after the implementation of the National Historic Preservation Act (1966), among other legislation actions passed around that time (Black and Jolly 2003). Geoarchaeology became an essential part of archaeological work, and the number of geoarchaeologists working in CRM projects increased greatly since the 1980s (Waters 1992). Today, CRM archaeology represents a great source of work for archaeologists and specialists associated with archaeological research, whose work has contributed to the development of methods and techniques, a matter that is still little known to many academic archaeologists (Schuldenrein 2007). Nonetheless, although originally seen with great concern by academic archaeologists, the practice of CRM has gained more and more acceptance (Holliday and Rotschild 2012).

One problem that had persisted in CRM archaeology has been the diffusion of its research, as most project reports were archived in an office and rarely finding their way into libraries or more diffused publication outlets – thus gaining the term "gray literature." More recently, however, such reports have been made public online as is the case of the Digital Archaeological Record (tdar.org). Furthermore, non-academic archaeologists are publishing their research in academic outlets and participating in academic congresses. In some projects, academic and non-academic archaeologists have begun to collaborate in joint projects and share more ideas and common issues of research.

Although still undervalued by many academic archaeologists, and geoarchaeologists, CRM work and non-academic work in general is of great significance as a record of sites that otherwise would not be studied and a source of methods and techniques applied to specific problems. Also, as the shrinking of budgets in academia continues, more and more jobs for archaeologists and geoarchaeologists are going to the non-academic sector. Therefore, the enhancement of collaboration between the two traditionally divided academic and non-academic field is necessary.

Emerging fields, focuses, and applications

Most of the development of geoarchaeology over the years is reflected in the process of technical specialization, which is constantly evolving into new fields targeting particular problems with specialized methods. These fields include, for example, experimental geoarchaeology, ethnogeoarchaeology, and forensic geoarchaeology, and 3-D and visualization geoarchaeology, among others. These fields, are shaping the academic and non-academic geoarchaeology in the twenty-first century and contributing to its advancement in academia.

On the topical side of geoarchaeology, other specialized fields are developing fast. Some include focuses on specific sites or settings, techniques, and regional issues. The case, for example, of underwater archaeology directed to studying prehistoric sites drowned by the post-glacial rise in sea level has gained importance in the study of paleoamerican geoarchaeology (e.g., CIT). Likewise, paleogeographic reconstructions of glacial landscapes have been undertaken in other parts of the world, particularly between the British Isles and continental Europe (Brown 2008).

Geoarchaeology as an environmental discipline

Many practitioners of geoarchaeology, including those with archaeological backgrounds, have come to realize the importance of the scope and extent of geoarchaeology in the context of environmental history and global environmental change (Butzer 2008, 2015; Brown, Bassell and Butzer 2011; Butzer and Endfield 2011).

In the study of past human societies and the environment, geoarchaeology has begun to establish alliances with other disciplines in an effort to integrate data to better understand a variety of global environmental issues that, although occurring in the past, have replications in the present. Some alliances with archaeology and environmental sciences have contributed to broader global environmental projects, such as IHOPE (Integrated History and Future of People on Earth), that aim more at understanding broad human-environmental problems (Costanza, Graumlich, and Steffen 2007). Beyond any contribution, such alliances are good as part of the previously mentioned need for advancement of the field.

Furthermore, although geoarchaeology focuses on the environmental context of humans, it is not limited to study the past. It can also study the present as means of understanding the past and understanding our human legacy in the future – the main tenets of the geoarchaeology of the contemporary past (Chapter 17). Beyond this intellectual value, geoarchaeological research of the contemporary past encompass a large number of applicable studies, such as those in forensics (Schuldenrein et al. 2017) and historical ecosystem restoration (Roos 2008; Balaguer et al. 2014), among others. Thus, the possibilities of applying geoarchaeology to environmental problems is one of the points in favor of making this field more visible.

2

Theoretical and Methodological Foundations

INTRODUCTION

In order to understand the nature of geoarchaeology, it is important to address general questions about its theory and method, topics that are scarce and virtually absent in the literature of recent decades, despite occasional debates and discussions in conferences, symposia, and forums. In contrast, in its early years, in the 1970s and early 1980s, a considerably large number of papers devoted more attention to theory and method, a trend that by the early 1990s had declined considerably (Leach 1992).

It is not clear why this trend to avoid theoretical aspects in the geoarchaeological literature has remained so persistent. It is possible that practitioners are taking theory for granted, but it is also possible that in current academic practice it is less common to write about theory in favor of technical papers. But based on the scarce sources in the broad geoarchaeological discourse and its parent disciplines, archaeology and the geosciences, it is possible to reconstruct the epistemological framework of geoarchaeology.

Thus, the main objective of this chapter is to address the fundamental questions regarding the nature of the geoarchaeological theory and method – if any exists. In so doing, it addresses, first, aspects of the philosophy of science and the contributions of the background disciplines (archaeology, geology, geomorphology, geography, soil science, and ecology) to the theoretical and methodological foundations of geoarchaeology.

THEORY IN GEOARCHAEOLOGY

Is there a unique way of reasoning in geoarchaeology?

Although there are multiple definitions of *theory* in philosophy dictionaries and treatises, it is possible to distinguish two conceptions of it: synthetic and semantic (Rosenberg 2005). The synthetic conception refers strictly to axiomatic statements that stretch from conjecture hypotheses to theory. The semantic conception refers to a set of theories, ideas, hypotheses, and other non-formal statements that support the structure of a discipline (e.g., archaeological theory). Therefore, the semantic conception of theory is the one that concerns the present discussion.

Theory comprises a series of models that guide scientists through the scientific process (Schiffer 1988), the superstructure of a science, and its methodology (Rhoads and Thorn 2011). In a simpler sense, theory represents the way scientists reason within their discipline (Von Engelhardt and Zimmerman 1988; Schiffer 1988; Baker 1999; Peregrine 2001). Accordingly, one can deduce that archaeological theory represents the way archaeologists reason, and geoscience theory, the way geoscientists reason. Thus, the question here is: do geoarchaeologists reason more like archaeologists or like geoscientists?

On the matter of geoarchaeological reasoning, Rapp, Jr. and Hill (2006) in their introduction (p. 2) and epilogue (p. 274) seem to imply that geoscience reasoning prevails in geoarchaeological work. However, given the diversity of academic backgrounds – practitioners may be trained in archaeology, geology, geography, or even in the biological sciences – it is difficult to generalize the geoscience-only view. Therefore, it is important to address the theoretical and methodological backgrounds involved in geoarchaeological research as a way to assess the diverse and common ways of scientific reasoning in its practice.

Theoretical models from archaeology and anthropology

Because geoarchaeology was born as a methodological necessity to solve archaeological research problems, archaeology, and thus anthropology, has undeniably influenced modern geoarchaeological thought and method. But here the problem becomes entangled with time, as archaeological theoretical models have changed through time.

The theoretical history of archaeology goes back to the early days of the discipline, particularly in the context of the first studies of prehistoric sites and the theories of evolution (Trigger, 2003). In the twentieth century, as archaeological excavations increased around the world, the theoretical models that explained the cultural traits and historical developments of the civilizations and peoples under study became the mode until about the 1950s (Trigger 2003; Renfrew and Bahn 2004). The development of archaeological theory saw an important impetus in the 1960s, initiating a period during which theoretical models multiplied (Schiffer 1988; Hodder 1992; Trigger 2003; Bentley and Maschner 2008).

One of the most influential theoretical models born after the 1960s was New Archaeology, which called for a more scientific archaeology, particularly in an environment where archaeology was dominated by the humanities. In general terms, the idea was to adopt the scientific method, which at the time was dominated by logical positivism. Those who adhered to New Archaeology were referred to as "processualists" because of the focus on process, as the basis for explaining the formation of the archaeological record. But not everyone agreed with the processualists. Thus, the "post-processualist" movement rejected the logical positivism of the New Archaeology (i.e., processual ideas), creating a series of currents that are difficult to define as a unified theoretical model (Fogelin 2007).

Despite the apparent incompatibility between the processual and post-processual currents of thought, in the end modern archaeology tends to blend useful ideas of both sides (Salmon 1992; Fogelin 2007). Thus, many of the ideas originally proposed by processualists (e.g., Watson, Leblanc, and Redman 1984, and the original 1971 edition) have created an impact difficult to erase from archaeology. On the other hand, post-processualist currents such as hermeneutics (Hodder 1992; 1993) and behavioral archaeology (Schiffer 2010) constitute an important part of modern archaeological thought.

Geoarchaeology has been for the most part on the margin of the processualist/post-processualist debate. However, the scientific boom experienced through the New Archaeology model led to the use of methods from the geosciences, which in turn made possible the creation of archaeological geology and geoarchaeology (Rapp, Jr. and Gifford 1982). Nonetheless, some post-processualist ideas are from time to time incorporated in some aspects of geoarchaeological interpretation, as is the case of phenomenology (see discussion in this chapter). In this sense, geoarchaeology, with its strong geoscience component, has evolved eclectically when it comes to incorporation of archaeological and anthropological models of thought.

Theoretical models from the geosciences

Geology has a longer theoretical history than archaeology. In fact, geology and the ideas of geological stratigraphy influenced the work of early archaeology, particularly by the geological stratigraphic interpretation schemes proposed by Baron Georges Cuvier (1769–1832) and Charles Lyell (1795–1875) (Rapp, Jr. and Gifford 1982). By the end of the 1800s, archaeology had moved away from the deep-time geological stratigraphic principles to focus on a smaller-scale cultural stratigraphy (Rapp, Jr. and Gifford 1982), but in the second half of the twentieth century, geology again became an important aspect solving certain archaeological problems, thus becoming the foundation of the nascent field of geoarchaeology (e.g., Renfrew 1976; Butzer 1976; Gladfelter 1977; Hassan 1979). In this process, in addition to geology, other geosciences joined the cooperation, particularly geomorphology, geography, and soil science, and many subdisciplines of geology itself, particularly sedimentology and stratigraphy.

In the context of landscape evolution, sedimentology, and soil science, geomorphology and physical geography contributed to the explanation of a series of archaeological questions about site formation and transformation processes, as well as the basis for the study of humans in the landscape and environmental changes. Of these disciplines, geomorphology has become important in the development of "landscape evolution models, which provide a context to human-environmental interactions" (Butzer 1982; Waters 1992; Frederick 2001).

Geography has contributed theoretical models to geoarchaeological research, particularly from physical geography, which technically includes geomorphology and soil geography, but also climatology and biogeography (Gregory 2000). Although not explicitly involved in geoarchaeological research, climatology and biogeography have contributed background models for the interpretation of environmental processes at various scales. Climatology and its branch paleoclimatology have contributed models that help explain phenomena such as the Indian Ocean Monsoon (IOM), El Nino Southern Oscillation (ENSO), and the North Atlantic Oscillation (NAO), and others, which have had impacts on modern and past environments. For the deep prehistoric past, models of atmosphere-ocean circulation models based on deep-sea and ice core data that explain phenomena relevant to human dynamics (e.g., marine isotopic stages, Heinrich events, the Younger Dryas, and the Bond events). Biogeography, a field with connections to ecology, has increasingly influenced the human-environmental approach in geoarchaeology (Chapter 5), particularly when long periods of time are involved in processes such as migrations, glacial refugia, land bridges, and other aspects that affect humans on long time scales (Chapters 6 and 7).

The other side of geography, human geography, has also contributed geoarchaeological interpretations through theoretical aspects of cultural ecology and human ecology, both of which have gained a strong foothold in anthropological archaeology, and certainly in the human-environmental tradition in geography. Cultural geography, a field straddling geography and anthropology, has been influential in the explanation of cultural-related phenomena in geoarchaeology. Independent from the ecological view, geography has contributed directly to archaeology and geoarchaeology, with theories of spatial analysis, some discussed in detail by Butzer (1982).

Theoretical models from other fields

Disciplines other than archaeology, anthropology, and the geosciences have also contributed theoretical frameworks to geoarchaeological research, but such influences are minimal and more focused on particular fields and approaches in geoarchaeology. Perhaps the most important of these disciplines is ecology, in both its biological sense and as a broader paradigm. The biological sense refers mainly to aspects of population, communities, and ecosystems, a matter that permeates geoarchaeological thought via the broader field of environmental archaeology. The broader ecological paradigm (discussed in Chapter 5) has contributed theoretical models developed in human ecology, cultural ecology, and historical ecology. Along with ecological theory, models from physics, statistics, cybernetics, and economics have been introduced. These include, for example, chaos and complexity theories, networks, and the increasingly popular panarchy models.

The social sciences and humanities have also contributed to certain aspects of geoarchaeological research. History has contributed theoretical models, particularly in areas of interpretations of civilization and their environments (Butzer 2012; Butzer and Endfield 2012) or in aspects linked to historical ecology, which is very important in the reconstruction of historical landscapes (Egan and Howell 2001).

THE GEOARCHAEOLOGICAL METHOD

Is there a geoarchaeological method?

If one browses the existing textbooks, treatises, and topical monographs on geoarchaeology, one can appreciate a variety of methodological models, some leaning more towards geology and geomorphology, some towards environmental archaeology, and others more towards broad human ecological aspects, experimental research, and some even to humanistic approaches. But despite this diversity, such methodologies conform to the "geoarchaeological method" at least in the sense of a unified field of research. Proof of that is that all these methodologies are applied in a series of steps (Figure 1.2), and that whatever the background and method, the practitioner works under the geoarchaeology umbrella.

The diversity of methodological approaches in geoarchaeology is in part the result of the range of academic and technical backgrounds of its practitioners and in part dictated by the needs of the different archaeological problems to solve. Therefore, it is valid to say that there is more that there is no archaeological method, but many methods adapted to problems. In the same way that theories in geoarchaeology originate from background sciences, particularly geosciences and archaeology, methodologies originate from the same sources. Thus, tracing the methodological models back to their origin is an important epistemological piece of information to better understand the nature of geoarchaeology. One can call these models methodological blueprints, because, as discussed below, they are adapted to the conditions and scales used in archaeology.

Geoscience methodological models in geoarchaeology

Geology has been the primary contributor of methods to geoarchaeology, particularly through the study of sediment and stratigraphy (Stein 2001a; Goldberg and MacPhail 2006), particularly through the adaptation of sedimentary and stratigraphic principles used in deep time (e.g., sedimentary rocks) to Quaternary unconsolidated sediments intertwined with cultural layers (Leach 1992; Waters 1992; Stein 2001a). Geology has also contributed methodological blueprints for a variety of other methods, including petrography and geochemistry, some of which have influenced methods to study soils and sediments (e.g., soil micromorphology).

Sedimentology and stratigraphy have been tools for the interpretation of the sedimentary record, especially through the use of *sedimentary facies* models, which although developed for reconstructing paleoenvironment in sedimentary rock sequences, have been adapted to recent (quatenary) depositional environments (Leach 1992; Brown 1997; Feibel 2013). Likewise, the adaptation of geological stratigraphic models to short time scales has been an important development in geoarchaeology, seen, for example, in the establishment of *allostratigraphic* units, as opposed to lithostratigraphic units designed for longer periods of time (Holliday 2001).

Geomorphology constitutes an important contributor of methodological blueprints to geoarchaeological research. Firstly, geomorphology provides tools to study processes linked to depositional environments and the evolution of the landscape. Secondly, geomorphology provides the tools to study features and landforms built or modified by humans (e.g., mounds, terraces, dams, dikes, and dikes, etc.), and human impacts on the landscape (e.g., erosion and deposition). Thirdly, geomorphology studies site settings, and processes that transform sites.

The study of soils in geoarchaeology has been carried out in conjunction with geomorphology, in a field called *soils geomorphology*, (Birkeland 1999; Holliday 2004), which has contributed various methods of interpretation of the geoarchaeological record at different levels. At a site level, soil science has provided methods and techniques to study site formation processes (Butzer 1982; Holliday and Mandel 2016). Many of these methods involve the study of soil nutrients as evidence human influences on soils (Holliday 2004). Nutrients have also been essential in non-soil deposits, particularly in the study of activity areas (Goldberg and MacPhail 2006). Soil micromorphology, a method that has its basis in petrography, has been an important methodological asset in geoarchaeology because it helps interpret soil processes at fine scales, as opposed to coarser scales used in soil interpretation at the level of landscape (Goldberg and MacPhail 2006; Garrison 2016).

Geography is a discipline that straddles the social and natural sciences, as the discipline is divided by the interconnected fields of physical and human geography. Physical geography as a whole is focused on the interaction between the spheres: lithosphere, atmosphere, hydrosphere, biosphere, and sometimes cryosphere. Furthermore, physical geography is often linked with human aspects, to the point that there is a recognizable cultural physical ecology (Gregory 2000). Perhaps the most important contributions of physical geography to geoarchaeology are through geomorphology, whose view, unlike that of geological geomorphology, focuses more on human time scales. Human geography, on the other hand, has contributed many methodological models to archaeology and geoarchaeology, particularly spatial analysis such as central place theory, the nearest neighbor analysis (Butzer 1982) and most important studies of the cultural landscape, which is the basis for studying archaeological landscapes (Wilkinson 2003).

Geographic spatial methods, which are often application of geo-technologies, are essential in archaeological and geoarchaeological research, through the use geographic information systems (GIS), environmental remote sensing, global positioning systems (GPS), and unmanned aerial science (UAS) in the recording, collection, storage, and analysis of spatial data. Arguably, however, many see them as techniques, but some proponents within geography see them more as *geospatial science* that is evolving into a systematic way of seen space, time, and distribution of phenomena in space and time.

Archaeological methodological models in geoarchaeology

Geoarchaeology appeared within the field of archaeology, that is to say, as the means of solving certain archaeological problems related to site formation, taphonomy, and context (Butzer 1982). Therefore, the influence of archaeological methodological models in geoarchaeological research is undeniable regardless of the research approach, tradition, or focus (Figure 1.1). In fact, many geoarchaeologists with geoscience backgrounds do not realize how much their work is influenced by archaeological and anthropological methods.

In North America, and in the Americas in general, the archaeological method has a strong influence from anthropology and its other branches, particularly ethnography, but due to the influence of the American literature across the world, the influence of anthropological archaeology is now seen in many studies across the globe.

The most important methodological model that archaeology has provided is the study of cultural stratigraphy, an aspect that in theory all geoarchaeologists should know. This means not only the interpretation, but also the understanding, of formation processes, some of which are explained by behavioral aspects. Perhaps the foundations of this influence lie in the study of the many concepts of behavioral archaeology by Michael Schiffer (1987, 2000) and archaeological stratigraphy by Harris (1989). Although the Harris Matrix sometimes does not explain certain stratigraphic problems in geoarchaeology, it is still widely understood by most practitioners (Goldberg and MacPhail 2006).

Other methodological models from archaeology came from the fields of experimental archaeology and ethnoarchaeology, both of which have helped create the subfields of experimental geoarchaeology and ethnogeoarchaeology (MacPhail 2016; Tsartsidou 2016). These studies, although mostly focused on site formation processes, are now being extended to understand processes of landscape and environmental change (see Chapter 16). In this process, the study of modern human-environmental processes, as they occur, or have occurred in the recent past, are also beginning to influence geoarchaeological methodologies (see Chapter 17).

Methodological models from other fields

There are numerous minor influences of methods in other sciences in geoarchaeology, of which perhaps the most important are those associated with ecology (in its pure form or via human ecology), which has influenced the study of archaeological landscapes and processes related to the ecological context (sensu Butzer 1982). In particular, historical ecology has contributed to studies of historical landscape transformation and reconstructions where information in the form of sedimentary archives and features are correlated with historical accounts and maps.

Quaternary paleoecology, a field that is part of biogeography but also claimed by environmental archaeology, has influenced geoarchaeology particularly in terms of sampling for studies of paleoenvironmental reconstruction and the idea of multi-scale interpretations. Quaternary paleoecology and geoarchaeology share so many theoretical and methodological aspects that often work very closely in study of human-environmental relations.

GEOARCHAEOLOGICAL MODELS OF INQUIRY AND INTERPRETATION

As implied in the previous sections, it is difficult to pinpoint a unique methodological model because of the diversity of models adapted to geoarchaeological research. Therefore, finding a single *modus operandi* for the practice of geoarchaeology is difficult if not impossible. Nonetheless, the literature provides prescribed models for particular problems, environments, scales, and research problems. Thus, notable contributions to methodological models in the past two decades include treatises such as those by Waters (1992); Brown (1997); Stein and Farrand (2001), French (2003); Holliday (2004), Goldberg and MacPhail (2006), Garrison (2016), and collected volumes by Goldberg and Holliday (2001); Pollard (2009), Brown, Bassell, and Butzer (2011), and Wilson (2011), as well as numerous special volumes in various journals.

One aspect that is evident in the literature is that sometimes theory and methods sometimes follow the focus and approaches in geoarchaeology (Figure 1.1). For example, the geomorphology and soils approach has a more *de rigueur* method, often involving cursory descriptions of soils and sediments in the field followed by laboratory analyses such as particle size distribution, texture, and bulk density, among others, (Stein 2001b; Holliday 2004; Goldberg and MacPhail 2006). In addition to field descriptions, geomorphological research in geoarchaeology uses the traditional tools of aerial photographs and topographic maps, which are increasingly being replaced by the use of satellite imagery, GPS capabilities, and combinations of photographic and GIS technologies for 3-D models (Ghilardi and Desruelles 2009; Kvamme 2016).

The archaeological geology approach follows several models that are more clearly explained in some of the textbooks focused on this approach (e.g., Herz and Garrison 1989;

Garrison 2016), applying methods developed in fields such as geochemistry and petrography, among others, to archaeological problems. Although aspects of geomorphology and soils are implicit in archaeological geology, the theory and methods are more akin to the general field of geology.

The human-environmental approach follows a more varied array of theories and methodologies, combining aspects of environmental archaeology, geography, and ecology, but with strong links to the geomorphological and soils traditions, particularly in aspects of the landscape. An example of the latter can be seen in some of the important monographs that represent this approach (e.g., Rosen 1986; Wilkinson 2003; Hill 2006; Cordova 2007) as well as numerous other papers in journals.

Having recognized the multiple theoretical and methodological influences on geoarchaeology from the background sciences, it is important then to discuss the philosophical aspects of the geoarchaeological research process, particularly those related to scientific reasoning and the scientific method, both of which are important for understanding the general influence of science in the practice of geoarchaeology.

The basis of scientific reasoning in geoarchaeology

Philosophers of science have strived to formulate rigorous structures and language to be used in all scientific fields. But because many of these formulas were created using physics as a model, many other sciences, and particularly social sciences, have either dismissed them or adapted them to their own circumstances (Rosenberg 2005). Nevertheless, the tenets of the philosophy of science, even in its most rigorous form – logical positivism – still matter in all sciences because they still lead the scientific process (Salmon and Salmon 1979; Holt 1982; Frodeman 1995).

The most common aspects of the scientific processes in philosophy of science are *explanation* and *prediction*, and the forms of inference used to achieve them. In a strict sense, explanation is a scientifically valid argument, enunciated in a logical structure that explains an event, including the *explanandum* (the phenomenon to be explained) and the *explanans* (the statements that explain the phenomenon) (Watson LeBlanc and Redman 1984; Rosenberg 2005; Von Engelhardt and Zimmerman 1988). Prediction, on the other hand, while also involving scientifically valid and logical expressed argument, is meant to envisage an event in the future based on observations.

Explanations and predictions are constructed through inference or reasoning, which can be deductive, inductive, or abductive. In general terms, deduction leads to a conclusion from a general rule through premises; induction leads to a rule from a case observed, and abduction chooses (abducts) the best of the cases observed to provide an explanation (Von Engelhardt and Zimmerman 1988).

In simpler terms, deduction leads to a necessary truth, induction to a likelihood, and abduction to the best possible likelihood. But according to the logical positivists, deduction is the only scientifically valid of the three ways of inference. Two of the great exponents of logical positivism, Carl Hempel and Paul Oppenheim, established the deductive-nomological model, in which all valid conclusions are based only on the laws of nature (i.e., the covering law model), which in turn establish the links between explanans and explanandum (Rosenberg 2005).

However, the deductive-nomological models contrast with many scientific fields, where the universal laws of nature do not apply, as is the case of archaeology, which is more concerned with the causes of phenomena and not universal laws (Hodder 1999). The reaction to logical positivism embraced by the supporters of the New Archaeology has received heavy criticism from the various post-processualist schools, each of which have proposed a number of alternative models (Fogelin 2007; Orser, Jr. 2015).

Reactions to the universal laws of natures are by no means inherent to social scientists. Even in natural sciences, such as geology, the universal laws of nature have received a significant dose of criticism (Von Engelhardt and Zimmerman 1988; Schumm 1991; Frodeman 1995; Baker 2000). Because geology is a historical science, that is, a science that studies cause-effect processes over time, the covering law model of the Hempel-Oppenheim school of logical positivism does not fully serve the purpose of causality (Baker 1999). Thus, geosciences have tended to replace the laws of nature by the cause-effect model as the link between explanans and explanandum (Inkpen 2004).

Abduction (retroduction) as the alternative form of reasoning to induction and deduction is more widely used in geosciences (Von Engelhardt and Zimmerman; Baker 1999, 2000; Inkpen 2008). The usefulness of abduction in the geosciences lies in that most of the data is obtained from incomplete records such as erosion remnants, stratigraphic records with hiatuses, and geochemical traces in sediments and organisms (Inkpen 2008). Interestingly, abduction is also a form reasoning applied to the interpretation of certain parts of the archaeological record, particularly in reconstructing features and artifacts out of only remains (Orser, Jr. 2015).

Another problem with logical positivism in geosciences is with the process of *prediction*, which in the strict sense of explaining future phenomena does not fit reasoning in geology, whose focus is on explaining past phenomena. Thus, the idea of prediction is adapted to the geology in the form of postdiction or retrodiction (Von Engelhardt and Zimmerman 1988). Thus, in view of the inadequacies of explanation, prediction, and the logical positivistic rigor of the deductive model, geologists have gone back to the proposals by American geologist Grove K. Gilbert (1843–1918), who proposed that geology had its own way of seeing the Earth, for which a method of its own is necessary (see Schumm 1991; Baker 2000).

Despite the criticism of the use of induction and deduction, these two forms of reasoning have an important role in a research project, particularly when data are scarce, and some sort of hypothetical model has to be developed (Holt 1982; Schumm 1991). Of interest in the case of the geosciences is the suggestion by Victor Baker (2000), for whom two types of reasoning style are present in the geoscience inquiry, one theory-directed and one earth-directed. The former imposes strict prescribed theoretical models on the study of the earth, while the latter is guided by the earth phenomena themselves, and applies first abduction before deduction and induction (Table 2.1). It is important to note that in geoarchaeology, for the most part, the *earth-driven method* is the prevailing one, as in most cases no laws or models can be applied a priori to a problem (Table 2.1).

Despite the conciliatory views between the strict and more adapted norms of scientific inquiry, two main school of thought exist in science, one that proposes the following of strict logical formulas of the logical positivism (the Analytical School) and another that proposes alternatives to them (the Continental School) (Frodeman 1994). This dichotomy has received attention in archaeology (see Watson 1971; Schiffer 1988; Hodder 1999;

Table 2.1. Comparison of reasoning styles in earth science (From Baker 2000, with modifications by the author)

	Theory-directed	**Earth-directed**
Basis	Define elements of nature (systems) capable of controlled study	Take the world nature "as it is"
Goal	Develop theories that explain Earth	Develop understanding of Earth
Emphasis	Idealizations: general principles presumed to apply at all times to all places	Real phenomena: real phenomena, concrete particular happenings, past and present
Characteristics	Experimental, predictive, mathematical	Experimental, historical, observational
Methods	Controlled experimentation and model simulation	Observation (in the field) to simulate hypothesis
Tools of study	Facts and theories	Signs
Role of "data"	Verification (validation) of model predictions	Signs providing indices of causal processes
Types of inference	Deductive analysis (rigorous and elegant) and inductive synthesis (for theory confirmation)	Retroductive (abductive) synthesis followed by deduction
Role of logic	Valid reasoning in regard to what we can say about Earth	Fruitful reasoning emphasizing what Earth says to us

among others) and in the geosciences (Von Engelhardt and Zimmermann 1982; Schumm 1991; Baker 2000; Inkpen 2003).

Among the many alternatives proposed by the Continental School is *hermeneutics*, a philosophy that in general terms refers to the interpretation of texts, although the idea has been expanded to the interpretation of the geological and archaeological records (Frodeman 1994; Hodder 1999). In geology, hermeneutics is a method that helps interpret sedimentary sequences, features in the landscape, and in general many aspects that are hidden in the record. In this sense, hermeneutics in geology has strongly influenced the interpretation of the geoarchaeological record.

In archaeology, hermeneutics involves the idea of context and origin, as well as aspects of reconstructive consciousness (Hodder 1992; Johnsen and Olsen 1992). These ideas have also influenced several aspects of the geoarchaeological record, particularly in matters concerning activity areas and transform processes in sites. Consequently, these ideas have also permeated the theory and method behind ethnogeoarchaeology and reconstructive models using 3D visualization.

Despite the arguments raised by those supporting the logical empiricist approach in science and those proposing alternative forms, most work in geoarchaeology still abides by the application of the scientific method. However, the impossibility of applying the scientific method to all cases of research sometimes require modifications using alternatives that better fit the needs of a particular type of research. Concerned about this problem, Stanley A. Schumm (1991) proposed several alternatives to be used in different cases of geomorphological research, some of which include the use of multiple hypotheses (see Schumm 1991: 13, 19–24). Notably, the use of multiple hypothesis fits better the case of reasoning by abduction, which as discussed above is more common in geology and consequently in geoarchaeology.

RECONSTRUCTING AND REPRODUCING THE PAST

Empirical approaches

Geology and archaeology, the parent sciences of geoarchaeology, are historical sciences in the sense that they aim at reconstructing processes and events of the past using field evidence that is often fragmented and in other cases invisible. This practice, however, contrasts with the way other sciences, such as physics, chemistry, and biology, whose research permits closer following of the empirical model prescribed by philosophers of science. However, empirical procedures are followed in certain forms of geoscience and geoarchaeological research, particularly when phenomena are observed in a modern context, but certainly not in the case when past phenomena are reconstructed.

Geomorphology, for example, is driven by the mutual interaction of two approaches, "dynamic geomorphology" and "historical geomorphology" (Cordova 2016b). Dynamic geomorphology focuses on present geomorphic processes, often monitoring them, measuring them, and putting them in all kind of contexts – tectonic, climatic, biological, and even human. Historical geomorphology, on the other hand, aims at reconstructing the processes of the past that led to the formation of present landforms, often using modern reference processes produced by dynamic geomorphology. In this sense, one can merge the idea of observed and historical processes under the characteristics of the earth-directed style to create something that one may call the *historical-observational method* (Table 2.1).

If one accepts the idea of a historical-observational method, then it is possible to say that this is the way most inferences are done in geomorphology and consequently in geoarchaeology. One can cite examples where phenomena, such as the formation of *anthrosols*, can be approached this way (see discussion in Chapter 12). In this case, it is implicit the idea that experimentation complements the reconstruction of the past, by linking observation with evidence of past phenomena, two aspects that only recently have begun to permeate geoarchaeological thought from the geosciences, particularly from soils geomorphology. Archaeology has also influenced some empirical aspects of geoarchaeology, particularly experimental archaeology and ethnoarchaeology (Chapter 17).

Geospatial technologies have also influenced some empirical aspects of geoarchaeology, particularly by replicating phenomena using computer models, where observed data are used as a basis to perform virtual experiments. In this sense, the idea of a controlled environment better fits the strict ideas of experimentation in physics and chemistry, where it is assumed that an environment is controlled.

Experimental and actualistic methods

The interest in observing natural and cultural site formation processes in the context of geoarchaeological problems led to the development of experimental geoarchaeology and ethnogeoarchaeology (MacPhail 2016; Tsartsidou 2016). The former focuses on the deliberate replication of human activities deduced from the geoarchaeological record. Such activities may include sources of raw material, soil transformation by farming, site abandonment and destruction, among others, with the purpose of observing processes. The latter also focuses on the same processes as experimental geoarchaeology, but as they occur organically in traditional societies (see Chapter 16). Both approaches have the purpose of establishing cause-effect relationships of phenomena related to the formation of the geoarchaeological record, which render them useful for reconstructing, if not hypothesizing, human-environmental relations in the past.

Despite its usefulness to answer some research questions, experimentation in archaeology and geoarchaeology have received criticism, particularly from the post-processualists, who criticize reasoning by analogy, a view that is seen as grounded in the logical positivism idea of scientific explanation (Fogelin 2007). Other sources of criticism come to the experimental method from practical issues, particularly with the problems involved in the assumptions made by the use of modern analogs (Goldberg and MacPhail 2006) (see discussion in Chapter 4).

Modeling in geoarchaeology

If one searches using the key words "modeling," "computer modeling," and "simulation model" in the webpage of *Geoarchaeology: An International Journal*, one finds a considerable number of articles that refer to research using modeling. Yet hardly anything is written on the epistemological aspects of modeling in geoarchaeology and their rationale as a methodological tool. Although a complicated semantics exists in defining models, it is important here to at least review the concept of model and its application in geoarchaeology.

In simple terms, models are abstractions of reality containing what are understood or believed to be salient features of that reality (Zimmermann and Artz 2006). This suggests that the modeler chooses the features of reality to model. However, the purpose of modeling beyond abstracting reality lies in the capability to make predictions and retrodictions about concepts, objects, systems, data, processes, and events (Sibley 2009, p. 255).

Although the purpose is understood, the process of modeling is a philosophical matter; according to theoretical biologist Robert Rosen (1991), a model is the means of communication between the real world and the formal world through encoding and decoding causal relations and inference. These relations entail processes of causation in the real world, where the phenomenon is studied, and a mode of inference or entailment in the world where the phenomenon is modeled. Under this scheme, a model permits predictions (or retrodictions) based on the mathematical inputs entered by the modeler, who also submits it to strict validation or verification, which in modern technological terms is carried out with the help of computer programs.

But not all models need mathematical inputs and validation, and not all models are generated by computers. Models can also be verbal, diagrammatic, and physical (Sibley 2009). Some models provide indices for a heuristic purpose, that is to say, simplify or abstract processes for means of communication as in the case of diagrams that represent complex processes and their relationships in an abstract for textbooks. In other cases, models become even more abstract as they ignore dimensions, as in flow charts that simplify cause-effect sequences, feedback loops, or aspects such as steps or processes.

Most models aim at interpreting complex aspects of the real world in a systematic sequential way, as in the Harris Matrix, whose purpose is to organize sediment layers and features in a chronological sequence in highly complex archaeological stratigraphic sections (Harris 1989). Various specialized computer methods used for arrangement of topographic and stratigraphic data into 3-D and 4-D models (Ghilardi and Desruelles 2009; Kvamme 2016). Computer programs have been also developed to systematically organize complex data, as numeric ages that need to be calibrated and sequenced with respect of depths (e.g., Bacon), as to create an "age model."

Attempts to solve the complexities of the world using diagrams occurred at several times during the second half of the twentieth century. In the 1960s the developers of the Systems Theory made popular the use of diagrams showing positive and negative feedback that allowed the mapping of cause-effect relations between phenomena – geomorphology, for example, used such diagrams.

In subsequent decades, the diagrammatic system theory models evolved into more complex models such as *chaos* and *catastrophe* theory models, which in turn gave rise to the concept of *panarchy* and resilience models, now popular in the social sciences, biology, and ecology (Holling 2001; Scheffer 2009) Many such models have gained ground in geoarchaeological research, particularly among the practitioners of the human-environmental tradition to explain complex processes involving nature and society. One example is the resilience model, which simplifies the complex world of variables between rise, decline, and resilience, connecting them with a *Möbius strip*.

Models in science essentially entail reasoning by analogy (Sibley 2009). The best example of analogy in geoscience is the *sedimentary facies model*, which is an analogy between the attributes of stratified deposits and modern known depositional environments (Reading 1986). The facies model is used in the interpretation of depositional environments of the past and organizing them in a sequence (time) over a certain area (space), using attributes of modes of sediment deposition (e.g., fluvial), thus serving the purpose of interpretation (heuristic model).

Increasingly important in the reconstruction of past processes are simulation models, whose design is based on analogical reasoning made functional with the help of computer programs. Such models, widely used in paleoclimatology, geomorphology, and other sciences of the past, have been increasingly developed in archaeology and geoarchaeology.

In the strict meaning of the word, prediction means modeling future events or trends. Simulating or reproducing process of the past would be retrodiction, a process that uses the same inputs and logics as prediction (Beven and Young 2013). It is for this reason that instead of "predictive model" the term *simulation model* should be used because of its neutrality between the past, the present, and the future.

Computer simulation can be considered a form of experimentation because it entails processes carried out in a controlled physical environment (Barberousse, Franceschelli, and Imbert 2009). In this case, however, the controlled environment is not a laboratory, but a virtual environment where inputs and factors involved in the process can be added, eliminated, or changed. Experiments with computer simulations are performed in cases when real physical processes cannot be reproduced either because of their time and spatial scales or because of their magnitudes.

In archaeology, site formation processes can be simulated, as is the case of formation of sediments in caves (e.g., Smyth and Quinn 2014), but also at a continental level paths of hominin migrations, which would be impossible to reproduce with the small number of dated sites (e.g., Carotenuto et al. 2016). In particular, 3-D and 4-D models generated through remote sensing techniques or developed in a GIS environment have become important as research planning and pedagogical tools (Challis and Howard 2006; Kvamme 2016). Needless to say, despite criticism, computer simulations have become an important methodological tool that is changing the way geoarchaeology studies archaeological and ecological contexts. The important aspect is to understand how simulation works and what its application is good for.

Phenomenology in geoarchaeology

In archaeology during survey and excavation, archaeologists reflect and visualize in their minds the landscapes they reconstruct out of the remains they report (Embree 1992). Similarly, the geologist does the same process when looking at outcrops, first envisioning past environments, in an initial process of cognition, and later turning them into the language of mathematics and geometry, making them both precise and verifiable (Frodeman 1995). In truth, experience and perception are two important phenomenological aspects that play an important role in field-based sciences whose purpose is to reconstruct the past, as is the case of geology and archaeology.

In a broad sense, phenomenology is a philosophical view that deals with the experiences that form people's views of the world. In archaeology, the phenomenological approach has been applied mainly to monuments in the field of landscape archaeology, often serving as a way to explain how its inhabitants experience the landscape (Owoc 2004; Tilley 2008) Although this idea is more developed in British landscape archaeology, it has important representatives in America, particularly in the ethnoarchaeological research (Shennan 2004). Nonetheless, the phenomenological approaches in archaeological interpretation has been criticized in the context of its subjectivity and the differences between modern and past social contexts (Shennan 2004; Johnson 2012).

Another problem with applying phenomenology to aspects of human-environmental reconstruction of the past is the bias that modern researchers have in an environment where science dominates reasoning, which raises awareness about ongoing phenomena. In contrast, however, inhabitants of Europe in the fifteenth century did not know they were living in a Little Ice Age, nor did their ancestors a few centuries earlier have the idea of a warming period. Likewise, Mayan farmers at the end of the Classic period were not aware of the broader context of environmental changes occurring at the moment.

However, phenomenological approaches in geoarchaeology and in general in environmental archaeology can be put in practice in what William Meyer and colleagues (1998) call *heuristic analogies*. "The analyst draws upon them selectively to develop a chain of ideas about climate-society relations, and valuable lessons may be learned from the process of drawing analogy" (Meyer et al. 1998: 220). Such views have been proposed as intellectual exercises and ways to grasp the complexity of past environments in geoarchaeological research (Bell 2007; Jusseret 2010; Cordova and Porter 2015), some of which can be useful in the classroom and outreach lectures to the general public.

THE EXPLANATION OF A COMPLEX AND CHAOTIC WORLD

Explaining processes and events that occurred in the past, as is the role of geology, archaeology and geoarchaeology, faces the problem of reconstructing a past that, like the present, is characterized by complexities and chaos. To solve the problem of understanding complex processes, mathematics created the theories of complexity and chaos, some of which have been adapted to physics, biology, ecology, and even the social sciences. Certainly, modified versions of these theoretical models have made it into archaeology and geoarchaeology.

One aspect of complexity in geoarchaeology comes with the analysis of processes at different scales, which has been discussed extensively in the geoarchaeological literature (Stein and Linse 1993; Rosen 2007a; Wilson 2011; Contreras 2016). The discussion on scale, or multi-scalar analysis, which includes what has been generally proposed as micro-, meso,- and macro-scales (1982), in which the micro-scale is understood as the focus on levels of artifact and site, useful for determining site formation processes.

The complexities found at a meso-scale involve the interpretation of phenomena in a given territory of research (basin, basins, or region defined by other boundaries, arbitrary, or natural) often refers to the evolution of the landscape, an aspect that still has ties with behavioral archaeology, but more with the background aspects such as climate and direct management of the land. Aspects of such nature can be seen in the series of studies dealing with the management of the Mayan rural landscape where analyses of the record constantly cross-cut different scales (e.g., Dunning and Beach 2010), where an extensive network of on-site and off-site research locations permit a broad understanding of complex processes across scales. But not all archaeological projects achieve these objectives, particularly when they are focused on one site (see discussion in Chapter 4).

At a macro-scale, that is a more continental or global scale, complexity refers to background phenomena such as global climate change. More recently these phenomena have addressed issues of atmospheric carbon and its effects in cooling (Ruddiman 2014). Although some meso-scale phenomena can be studied separately, in more recent decades the trend has been to place them in the macro-regional or global context. Then, studies of the Paleolithic refer to marine isotopic stages or events such as the Younger Dryas and the 8.2 ka event, while others on cyclic or recurrent phenomena such as ENSO (El Niño Southern Oscillation), NAO (North Atlantic Oscillation), and IOM (Indian Ocean Monsoon), among others. But the complexity lies on the causal explanation of the changes at a meso-scale. Are changes in the cultures of the Americas really explained by the recurrence of El Niño? This question is tricky sometimes even with hard evidence of say flooding or drought. Attributing weather phenomena as major causes to the decline of a civilization may be a simplistic explanation of a causality that involves complexity of interactions between natural and social phenomena. In this case, perhaps before invoking macro-scale phenomena, it would be better to determine some of the meso-scale controls of climate and environmental change.

CONCLUDING REMARKS

The brief epistemological review of this chapter is meant to review the foundations of theory, method, reasoning, and knowledge in geoarchaeology. This review suggest that this field is strongly influenced by theoretical developments in both archaeology and the geosciences. However, despite being born out of a need to address archaeological problems, the geosciences dominate in terms of theoretical and methodological foundations. But, unlike the ordinary practice of geosciences, geoarchaeology acts in a cultural context and works in tandem with the archaeological process (Figure 1.2). Therefore, archaeology and anthropology provide some theoretical basis for interpretation and some methodological aspects for research in archaeological contexts.

Although the geoscience-archaeology relationship is understood, at least on paper, the it is a complex one in practice, because geoscientists and anthropological archaeologists are interested in different kinds of theories. As Kent Flannery puts it, natural science theories aspire to explain "how" things occurred, while social science theory "why" they occurred (Flannery 1986: 3–4). Therefore, for the sake of mutual understanding, it is important that practitioners of geoarchaeology, whatever their academic background, be aware of theories and ways of reasoning on both sides of the relationship.

3

The Geoarchaeological Record

Concept and Contexts

AN EPISTEMOLOGICAL BACKGROUND

The geological and archaeological records as a background

In the geoarchaeological literature there are frequent references to "the record," presumably implying the geoarchaeological record. One can search the literature, dictionaries, and encyclopedias and there is no definition or discussion of what the geoarchaeological record is or includes. In contrast, the historical, archaeological, and geological records have received more attention in their respective fields.

In the absence of any definition of geoarchaeological record, one must search its meaning in at least four sources: first, in the concept of concept of *record* in in geology and archaeology, the parent disciplines of geoarchaeology; second, in the definition of geoarchaeology; third, in the practice of geoarchaeology itself; and fourth, in the usage of the term in the geoarchaeological discourse.

The geological record seems to be a simple one, to the point that not all dictionaries and treatises define it. One definition, for example, describes it as "The 'documents' or 'archives' of the history of the earth, represented by bedrock, regolith, and the earth's morphology: the rocks and the accessible solid part of the earth. Also, [it includes] the geologic history based on the record" (Bates and Jackson 1984, 204). Thus, one can assume that the geological record encompasses all rocks, unconsolidated sediments, and fossils of any age from Precambrian to the present.

The archaeological record, on the other hand, has been conceptualized in more complicated ways. Its meaning has evolved through time and in some conceptions, the record includes even more than what it is out there for the archaeologist to dig up. To understand this idea, one has to consider, first, that archaeologists not only look at artifacts but also at features (non-movable objects), ecofacts (natural materials used by humans), and sediments surrounding these materials, as well as materials that have disappeared or been removed from the record but that we know existed either because of traces left or because they are mentioned in historical documents (Peregrine 2001; Orser, Jr. 2015).

Additionally, archaeologists are interested not only in physical objects but also in aspects of human behavior, social structures, and economic relations that sometimes have to be deduced from the physical materials of the record (Orser, Jr. 2015). Therefore, defining the archaeological record is a complicated matter that is also linked to the different schools of thought in archaeology (Trigger 2006).

Prior to the rise of the New Archaeology, the archaeological record was seen as seriations of artifacts (Orser, Jr. 2015), a conception that was later severely criticized by the representatives of the New Archaeology because of its lack of functional arguments and linkages between static and dynamic contents (Binford 1968). Obviously, these arguments refer to the existing or retrievable evidence (positive evidence), but not to the invisible and missing evidence (negative evidence) (Hodder 1992; Lucas 2008, 2012). The negative evidence refers essentially to the parts of the record that have been removed by differential preservation or missed by the lack of proper sampling and recording necessary for the interpretation of material culture (Lucas 2012).

In view of the positive and negative evidence, there can be several meanings or interpretations of the record. Linda Patrik (1985) proposed five meanings, which in turn can be understood by parts of the record (Table 3.1). In turn, one can see a broader view of the record (meanings 1 to 5) and a narrower view (number 5 only). Gavin Lucas (2012) condensed the ideas by Patrik and others into only three meanings (Table 3.1, right). In Lucas's view, the archaeological record, or what he also calls the "total record," corresponds only to meanings 3, 4, and 5 in Patrik's scheme. From these schemes, one can infer that the archaeological record corresponds to what the archaeologist encounters in the field and sources at present, and what goes into field notes, drawings, photographs, or anything produced in the field, which in turn are the basis for interpretation of the archaeological record itself.

In view of the geological and archaeological conceptions, it may be assumed that the geoarchaeological record may overlap parts of the geological and the geoarchaeological record. The problem with this assumption is that geoarchaeology encompasses also theoretical and methodological aspects of geography, ecology, and other fields that conform broad range of geoarchaeological approaches.

In theory, one can deduce the geoarchaeological record from the definition of geoarchaeology itself, but this is complicated because the definitions of geoarchaeology vary through the literature (e.g., Gladfelter 1977, 1981; Hassan 1979; Butzer 1982; Rapp, Jr. and Gifford 1982; Leach 1992; Stein 2001a; 2001b; Rapp, Jr. and Hill 2006; Butzer 2008; Wilson 2011; Cannell 2012). In most cases the definitions allude to the geosciences as the active part and archaeology as the passive part (Canti 2001). But from the conclusion of Chapter 2, it is apparent that despite being dominated by theoretical and methodological aspects of the geosciences, geoarchaeology incorporates many methodological aspects of archaeology, and to a lesser extent from other fields (e.g., ecology). Therefore, the geoarchaeological record is unique, in the sense that it incorporates aspects of interest to the goals of geoarchaeology itself.

But beyond the definitions of geoarchaeology, opinions of the concept and extent of the geoarchaeological record may vary depending on the approach, i.e., the three traditions of geoarchaeology. Furthermore, the record may be perceived differently in the different subject areas of geoarchaeological research. For example, geoarchaeologists working in Paleolithic research deal with different parts of the record than, say, those working in historical archaeology. Another aspect that defines the way practitioners approach the geoarchaeological record is the use of different scales, which in turn defines the context at which processes and events are analyzed (Butzer 1982; Leach 1992; Stein and Linse 1993).

Table 3.1. The archaeological record meanings by Patrik (1985) and Lucas (2012), Source: Lucas (2012)

Fivefold Division (Patrik 1985)	Threefold Division (Lucas 2012)
Past objects and events	Artifacts and material culture
Material deposits	Residues and formation theory
Material remains	
Archaeological sample	Sources and fieldwork
Archaeological record	

It is also important to consider that conceptions of a geoarchaeological record may vary according to each practitioner's academic background and field of specialization (Figures 1.1 and 1.2). Therefore, the conceptions of geoarchaeological record may vary considerably from a minimalist perspective narrowed by specialization and focus to a maximalist, all-inclusive perspective that encompasses the diversity of academic backgrounds, methodological approaches, and scales of research. The latter, however, is the one that is henceforth discussed.

AN ALL-INCLUSIVE GEOARCHAEOLOGICAL RECORD

In view of the epistemological points delineated above, it is apparent that the geoarchaeological record should be formed by all types of evidence that satisfy the wide range of research objectives and the diversity of research approaches in geoarchaeology. Moreover, the record should include all possible evidence, positive and negative, as well as parts that are not obtained in the field but help achieve the goals of the discipline. Thus, the categorization of its parts should follow a range of variables from the more visible to the less obvious and missing (Table 3.2).

The *visible* part of geoarchaeological record comprises artifacts, features, sedimentary layers with all their visible depositional features, landforms, biological remains (including ecofacts), and all material evidence that can be recorded macroscopically (Table 3.2). The *invisible* part of the record includes three sub-parts: visible parts that are concealed; microscopic objects; and traces in the form of chemical residues (e.g., phosphorous in soils,

Table 3.2. Proposed divisions of the geoarchaeological record

1. Visible: tangible objects, features, sediments, etc.
2. Invisible:
   Concealed – visible structures buried or not exposed.
   Traces – chemical or physical traces of objects, processes or activities, or any other cryptic evidence that needs instrumentation to be detected.
3. Absent – inferred objects, features, processes, with no tangible, visible, or cryptic expression. Missing parts in the record.
4. Virtual – indirect evidence in the form of field notes, maps, publications, computer files.

and radiocarbon age determinations), physical imprints (e.g., footprints), chemical residues (e.g., residues of a particular activity in the soil), and any other cryptic form that requires instrumentation to be detected. The *absent* or missing parts in the record are not visible or detected, but they are inferred as having existed. The *virtual* part of the record, which includes anything recorded as notes, photographs, computer files, knowledge, informants' accounts, and oral and written history.

The parts of the record, whether visible, invisible, absent, or virtual, can be further classified in terms of their importance to the main objectives of the research, thus they can have a core, tangential, or background importance. The *core* part of the record includes artifacts, features, and sediments that are directly excavated or reported. The *tangential* parts of the record include also materials that are not central to the research but can be useful, such as additional information on the environment or data from a nearby site.

The *background* includes all the environmental information that helps with the interpretation. Some background information may be in the form of data or publications from previous research at the site or the overall region. The background information may also include notes, publications, and computer files of previous work at the particular site or locale of the research.

In addition to their importance in the research or focus, the parts of the record can be classified depending on their scalability. Thus, the *uni-scalar* ones are limited by the scale at which they can be studied (e.g., an artifact, a floor, and any feature with importance only at the site level), while the *multi-scalar* can be analyzed at various scales (e.g., a recurrent feature in the landscape, a volcanic ash marker, or a particular soil or feature formed by way of a climatic event). The analysis of multi-scalar components then varies according to the contextual level, a matter that deserves discussion when establishing the scale at which sampling and interpretation is to be performed.

The question as to what is the time limit in the geoarchaeological record should be defined in terms of the objectives of each study. While the geological record covers the entire history of the planet, the archaeological record only that time where humans and their hominin ancestors are present. Where, then, would the geoarchaeological record have its lower limit? This question is better addressed in Chapter 7, where situations referred to gray zones represent cases where the time frame of the geoarchaeological record antecedes that of the archaeological record.

Contextual levels in the geoarchaeological record

The *ecological context*, as conceptualized by K. W. Butzer (1976, 1982) and B. Gladfelter (1977), includes not only Shiffer's (1971) *systematic* and *archaeological contexts*, which basically refers to artifacts in a site, but also the context at higher levels beyond the archaeological site. Thus the idea of ecological context means not only putting artifacts in a deposit or the context of the sites and their formation processes, but also in the context of geomorphological and sedimentary settings, evolution of the landscape, and the record of biotic, climatic, hydrologic, and geomorphic influences at different scales. Such a multi-scalar contextual approach means that processes can be analyzed in the context of the site, the setting (sedimentary and/or geomorphic), the landscape, and the environment (Figure 3.1).

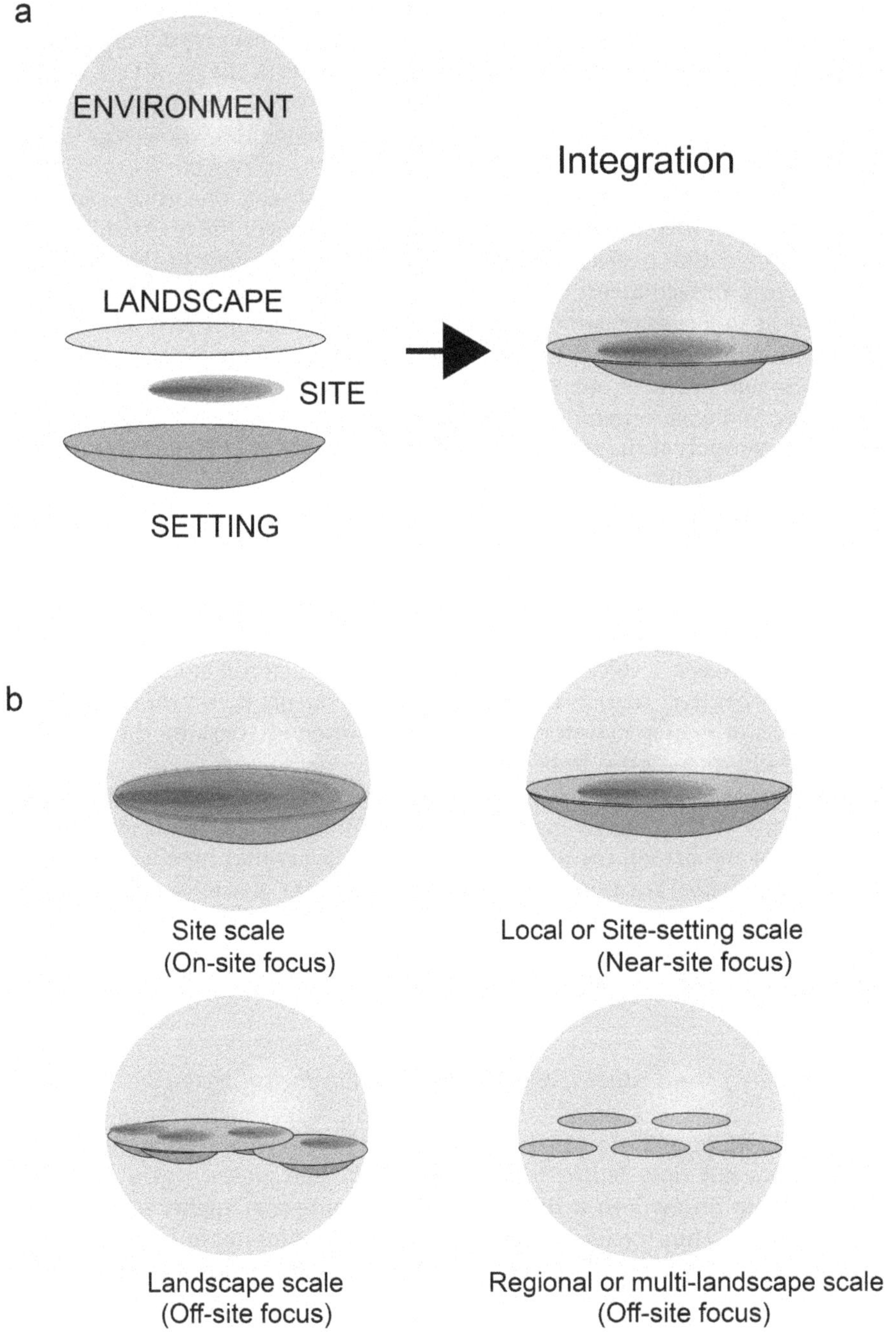

Figure 3.1. a) The four contextual levels in geoarchaeological research and the relations among them; b) examples of different levels of abstraction where the environment determines the scale of lower contextual levels.

Where are these four contextual levels from? The four contextual levels defined above were discussed in in the early geoarchaeology works (e.g., Fedele 1976; Hassan 1979; Gladfelter 1977; and Butzer 1982) as levels that define the space-and-time frameworks of every work in geoarchaeology. Besides, almost every paper in geoarchaeology mentions at least one of the four, or all four, depending on the research focus. Thus, the focuses of geoarchaeological research are varied, but one way or another, they pertain to one or more contextual levels (Figure 3.2). Each level relates differently to the archaeological and

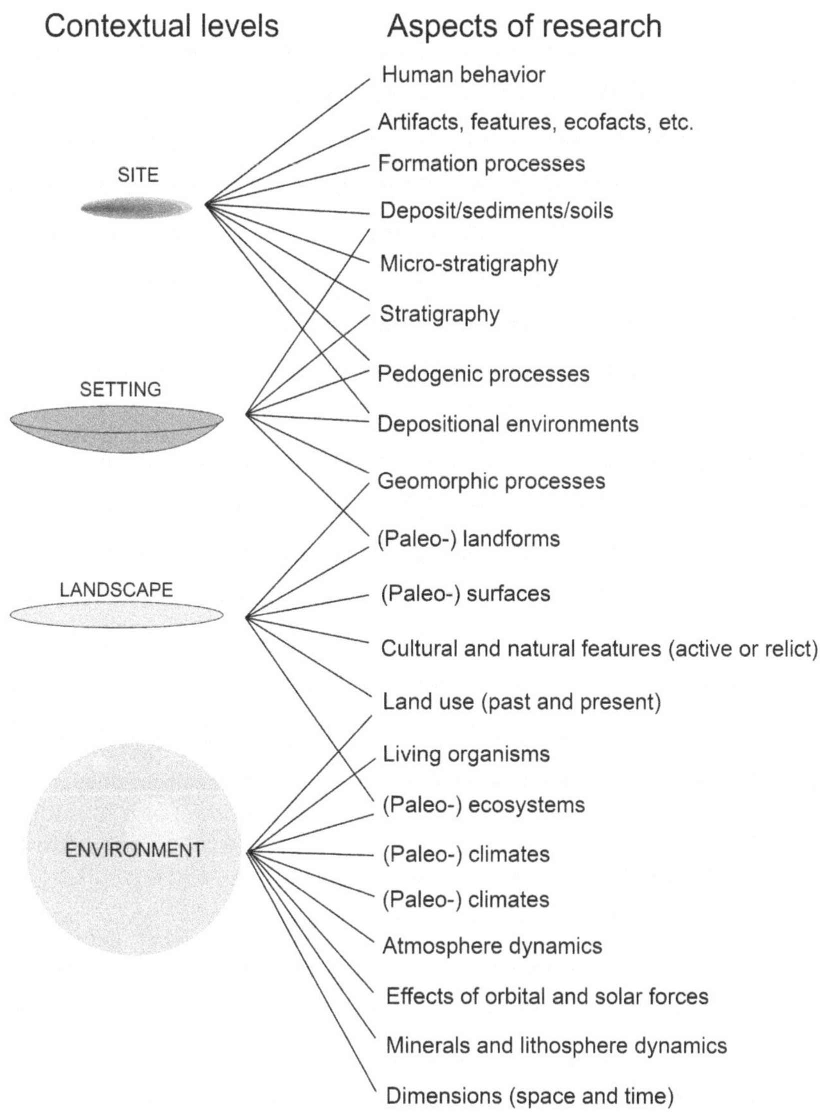

Figure 3.2. Schematic diagram of the four contextual levels and the aspects of geoarchaeological research they encompass.

Table 3.3. The four contextual levels of geoarchaeological research in relation to the archaeological and ecological contexts and their scalability dependence

Level	Archaeological Context	Ecological Context	Scalability*
Artifact	Yes	Yes	Not applicable
Site	Yes	Yes	Dependent
Setting	Partially	Yes	Dependent
Landscape	No	Yes	Dependent
Environment	No	Yes	Independent

*See Figure 3.1b for scalability dependence.

ecological contexts, and each has different properties, particularly regarding scalability (Table 3.3).

THE CONTEXTUAL LEVELS: SITE, SETTING, LANDSCAPE, AND ENVIRONMENT

Site: definition, boundaries, and properties

The *site* (i.e., the archaeological site) has been the main focus of geoarchaeological research from the beginning of archaeology as an organized field of inquiry, at a time when the main concern was to study site formation processes (e.g., Renfrew 1976; Fedele 1976; Gladfelter 1977). But at the same time, the idea of formation processes was seen as an extension of the depositional environment and geomorphic position (Gladfelter 1977; Hassan 1979), which in turn linked the site with its context, namely the landscape and the environment (Butzer 1976).

An archaeological site is broadly defined as "any place where physical remains of human activity exist" (Society for American Archaeology 2017). This definition implies that a variety of localities can be considered a site, from a small locality containing burned rocks and hearths to large buildings. The definition also implies that the presence of artifacts is not necessary to consider an area a site, thus an area with chemical residues or a rock wall with paintings or carvings can be considered a site. It is important also to notice that the definition above does not specify time in the past, that is to say, a human activity may have occurred tens of thousands of years ago, or only a few years ago, and it is still, by definition, an archaeological site.

The definition above, which is not much different from other definitions, leaves archaeological sites without proper physical boundaries. This not only creates a huge diversity of localities that potentially are sites, but also poses some problems in terms of assessing various geoarchaeological aspects of sites and non-site areas, which in turn affects the limits between on-site and off-site work.

The problem of site boundaries is difficult to assess in sites in circumstances of poor preservation, post-depositional transformation, and visibility. In cases of surveys in southern Europe (Bintliff 2000), Mesoamerica (Sanders, Parsons, and Santley 2012), and Australia (Rhoads 1992; Holdaway and Fanning 2014), a number of models have been tested to define sites and non-site areas in surveys. But there seems to be no "best" method, particularly in circumstances where scatters of artifacts occupy large areas, which in some cases forces archaeologists to use "mathematical" methods (Bintliff 2000).

Therefore, from a mathematical or statistical point of view one should imagine the limit between the site and non-site as a transition represented by a model using figurative isolines laid across the landscape from a core area where a particular activity of the past was more intense to an area where that activity is not perceptible. The non-site areas, or the areas where the isolines fade into nothing, can be either an area where the site has been erased by erosion or an area where there was no activity. In essence, the area of transition between the on-site and off-site area would constitute the *near-site* area.

The terms site and non-site are sometimes used in the literature interchangeably with on-site and off-site. However, a clear look at the use of terms should indicate that site and non-site are nouns referring to areas, while on-site and off-site are adjectives that are, or should be, used to define actions (e.g., off-site reconnaissance and mapping), and not necessarily the presence or absence of sites. One problem occurs, however, when the term non-site is used arbitrarily to define an area that archaeologists are not looking at or not expecting to find artifacts (see endnote discussion in Borejsza and Joyce 2017). In other cases, work in areas of features belonging to agricultural landscapes, e.g., agricultural terraces, is considered off-site (e.g., Goodman-Elgar 2008a), despite the fact that such features indicate human activity. Thus, how archaeologists perceive the non-site realms is sometimes arbitrary and concurrent with particular research strategies.

But beyond the problem of semantics and grammar, the issue refers to defining activity areas (Rhoads 1991). This questions the idea of whether a site is an archaeological site because artifacts are present or because a human activity occurred there (Foley 1981; Bintliff 2000). It is known that not all areas of artifacts are sites, which is a reason why the term *incidental site* is often used to refer to areas with artifacts in secondary context -usually not getting there as a direct effect of a human activity (Borejsza and Joyce 2017).

Quite often surveys report several sites that may have been one original single site that has been fragmented by erosion. Although this can happen in cases of Pleistocene open sites in uplands, due to long-time exposure to erosion, it can also happen in late Holocene sites where processes of erosion have been more dramatic dismembering original continuous sites into smaller units perceived as several sites (Figure 3.3).

An example of such a situation is seen in the Texcoco Region in Mexico (Cordova and Parsons 1997; Cordova 2017), where erosion in the uplands has dismembered remains of Aztec dispersed villages into small areas sometimes as small as erosion pedestals (Figure 3.4a). This particular location is indicated on Figure 3.3 as in between an area reported as a site. A study of the top soils in the soil remnants are the same age as the sites reported in the two adjacent sites (Figure 3.4b–c).

The problem of defining site limits requires also determining the transformation processes that have affected the site since its occupation, which does not necessarily involve the site itself or its immediate surrounding, but its broad geomorphic context. Taking the case discussed above (Figure 3.3 and 3.4), the broader geomorphic context reveals that as a result of such a massive removal of soils and sediment in the uplands, rapid alluvial aggradation downstream that buries sites of the same age (Figure 3.4d). Ironically, the buried sites do not show on the survey maps, which at the time of survey were not yet exposed but buried by recent alluvium. Thus, aspects of perspective,

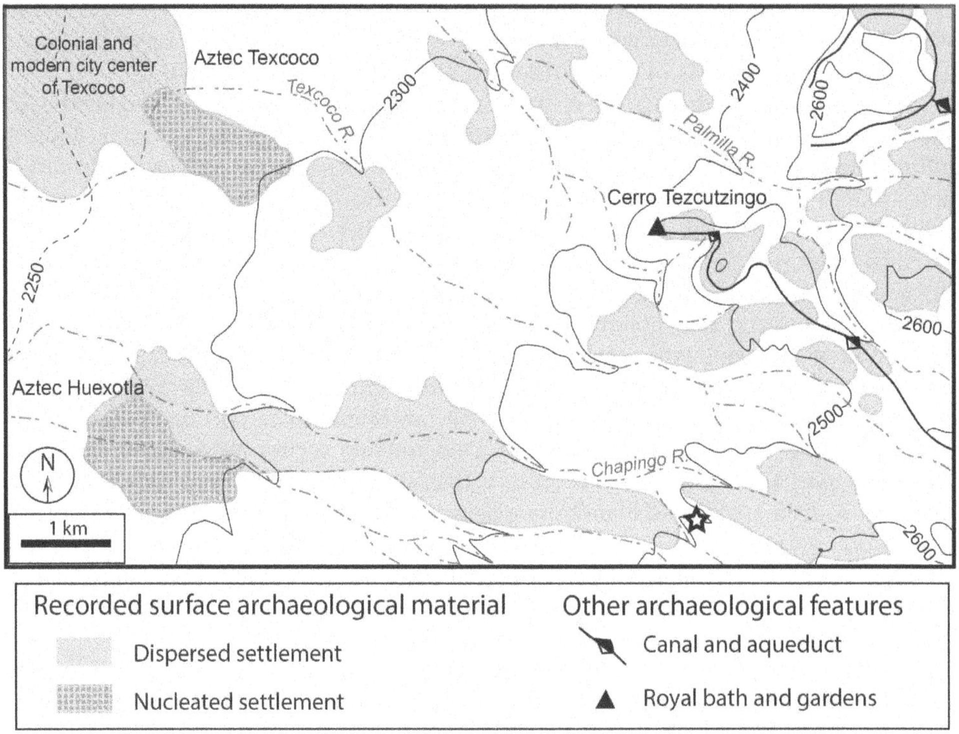

Figure 3.3. Aztec sites in Texcocan Piedmont, Mexico, recorded in survey (Parsons 1978; Cordova and Parsons 1997). The star indicates the location of the locality shown in Figure 3.4a.

particularly as seeing sites as surfaces can modify the idea of a site, which in turn should be seen as a three-dimensional object.

Perspective plays also a role in defining the site and non-site (off-site areas). If one takes the example on Figure 3.3, the sites depicted are based on the abundance of Late Aztec (1350–1550 CE), ceramics. The blank areas around the site are not off-site because they contain ceramics and structures of older periods, except for the eroded areas. This relativity of defining the dimensions of a site are best understood when the two next contextual levels are analyzed, that is to say, setting and landscape.

The setting: geomorphic and depositional

The *setting* involves all the matters that concern the static geomorphic situation (e.g., hilltop, valley, slope, cave, rock shelter, etc.) and the dynamic geomorphic processes associated with the formation and transformation of an archaeological site. Settings can vary depending on the geomorphic processes at the time of the occupation and afterwards: alluvial, eolian, lacustrine, palustrine, coastal, etc. For this reason, the setting should be seen as a geomorphic and/or depositional setting, but hereafter is referred to simply as setting.

Figure 3.4. (a) Pedestals at site Tx-78 in the piedmont of Texcoco, Mexico. (b-c) The topsoils on the pedestal contain Aztec III-IV pottery corresponding roughly to the fifteenth and sixteenth centuries, after Cordova (2017); (d) sediment exposure in the Texcocan plains with the following features: 1 = overbank and crevasse splay sediments, 2 = wetland sediment with Aztec III-IV pottery, and 3 = channel and overbank sediments burying older Classic and Preclassic paleochannels and sites, after Cordova (2017).

Because of its importance in understanding the context of the artifacts and the site, as well as the site formation processes, in geoarchaeological research the setting is a primary focus in explaining the context of sites (e.g., Waters 1992; Stein 2001a; Rapp, Jr. and Hill 2006; Goldberg and MacPhail 2006; Garrison 2016). Furthermore, the geomorphic and depositional setting is the connection between the site and the rest of the landscape. For example, the alluvial setting of the buried sites depicted in Figure 3.4d explains not only the formation and transformation of the site, but also their relation to the landscape – the very reason why the sites were not recorded through archaeological surveys.

The geoarchaeological setting explains also the preferences or the reason for the existing of and occupation or settlement at a particular location. Location in terms of certain activities, whether hunting, farming, habitation, or even defensive purposes, are often linked to the setting, which suggests that beyond the geomorphic and depositional dynamics, the cultural aspect is to be considered when studying site settings. However, a setting in the present landscape may not be the same as that of the time of occupation, if the geomorphological landscape has changed.

The term paleolandform suggests that there is a disparity between modern and past geomorphic settings. Because there is no term to define the disparity between modern and the original setting of an occupation, hereafter the term *conformable* and *unconformable* settings are used. In stratigraphy, the term unconformity is used to designate a discontinuity in the depositional sequence of a stratigraphic column, but deliberately here is applied to mean a discontinuity between modern and past settings, considering that the latter is the setting of an archaeological occupation or site.

The case of Figure 3.5 represents a model of sedimentary architecture that corresponds to the setting shown in Figure 3.4d. The floodplain setting, particularly in an overbank and close-to-channel position is still apparent. The stream does not flood the area in the same way, because it has been controlled, but technically floods can still and have occurred until recently. In this case the setting of the sites in the sediment and the modern setting is the same, that is, they are *conformable*. In contrast, the occupations on top of the pedestals showing in Figures 3.4a–c occupied a setting, probably a soil in a wide, flat-bottomed valley that is remarkably different from the present badland and erosive pedestal setting. In this case, the relation between the two settings is unconformable.

The interesting aspect of the setting unconformity explained above is that it is often implied that age contributes to such an unconformity. Certainly, as discussed in Chapter 6, the settings of early hominin occupations in the Pliocene and early Pleistocene are unconformable with their modern settings. However, setting unconformity is not exclusively tied with age. As shown in the case of Figure 3.4a, unconformable settings can occur when geomorphic processes accelerate. For this reason, judging setting unconformity in relation to age should be taken with care.

The landscape: its concept and properties

The concept of *landscape* varies depending on the different research fields in the sciences and the humanities. Therefore, it is not surprising that the meaning of *landscape* varies even among those disciplines associated with geoarchaeology, as is the case in archaeology,

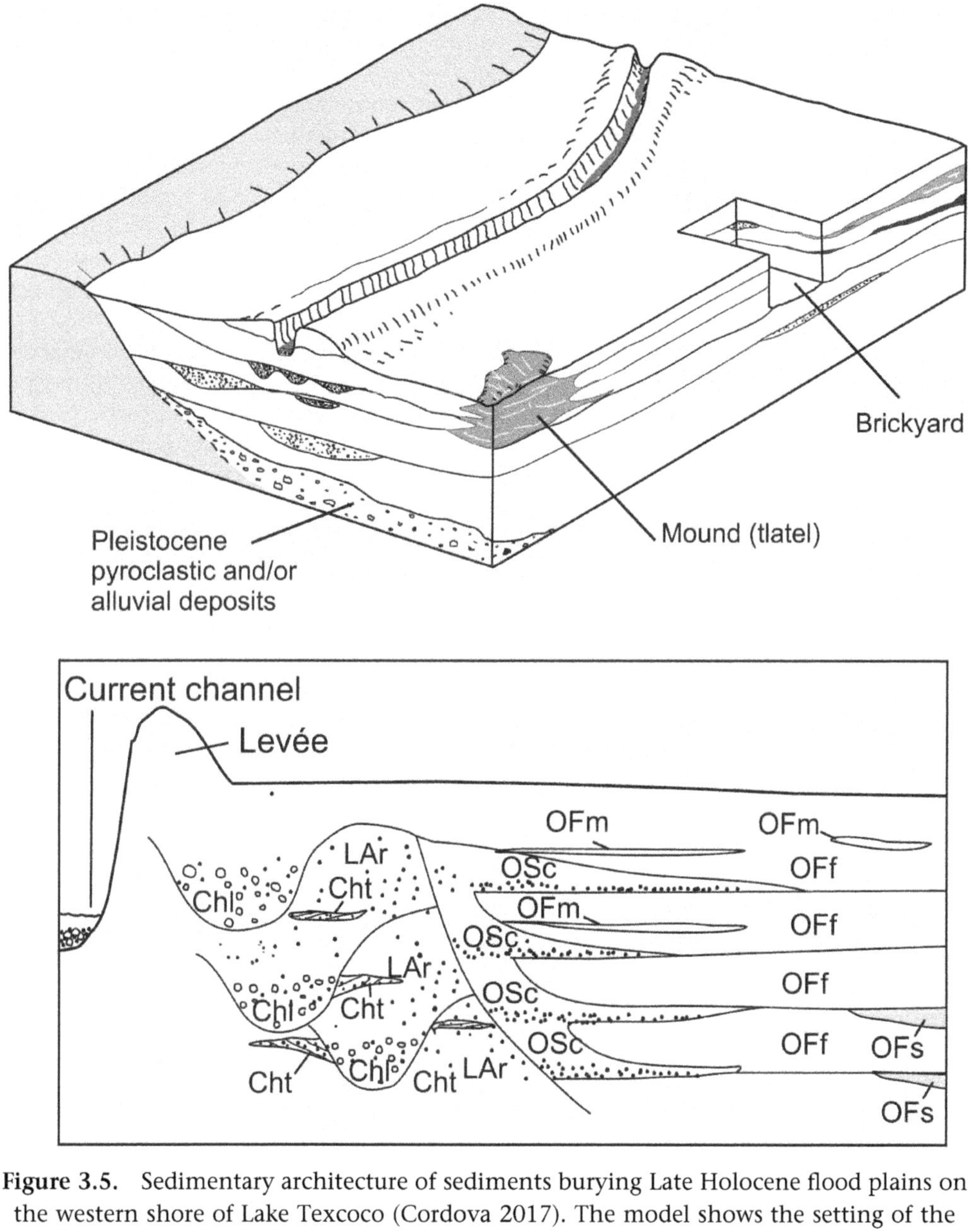

Figure 3.5. Sedimentary architecture of sediments burying Late Holocene flood plains on the western shore of Lake Texcoco (Cordova 2017). The model shows the setting of the locality in Figure 3.4d, which is exposed in a brickyard mine.

geography, geomorphology, and historical ecology (Wilkinson 2004; David and Thomas 2008; Bubenzer 2009). Furthermore, even within these disciplines conceptions may vary depending on the subfield and current thought.

In archaeology the concept of landscape can vary considerably depending on the focus applied to the interpretation of material culture. Thus, it can be a a focus on natural and cultural features (Wilkinson 2004), as the the expression of human cognition and activities (Hassan 2004; Strang 2008), or even as a meaning of perceptions and symbolism (Owoc 2004; Tilley 2008), among others. Of these, perhaps the most influential in geoarchaeology is the one conceptualized by T.J. Wilkinson (2003, 2004), because it relates vestiges of natural and cultural features to both the evolution of culture and the geomorphic evolution of the landscape.

In geography there are several meanings of the landscape, which can be summarized into two views: the holistic view, which was in vogue during the nineteenth and most of the twentieth centuries, and the structural view, which is particularly used particularly in physical geography, landscape ecology, and in the field of GIS science (Bubenzer 2009). While the holistic perceives the landscape as an amalgamation of features, the structural view conceives it as elements detachable from the whole landscape. This conception can be seen in the applications of Geographic Information Systems for which the management of layers implies that each layer contains at least one component of the landscape (e.g., geology, soils, land use, vegetation, hydrology). This in turn represents a resource to interpret land attributes (e.g., landform, soil, and vegetation), in the study of landscapes in geoarchaeology (Stafford 1995).

In geomorphology, the term landscape is often used to refer to the topography of a portion of the Earth's surface, often conveying the idea of evolution and change, as is often the case of the term landscape evolution (Schumm 1991). The idea is also focused on the morphologies of particular geomorphic environments (e.g., fluvial, eolian, coastal, karstic, volcanic) (Butzer 1976b). The meaning of the term landscape evolution in the geomorphological sense can often be used in geoarchaeology, particularly in the explanation of geomorphic changes, but often it appears as landscape evolution or with an adjective referring to the geomorphic environments (e.g., fluvial landscape).

In historical ecology, a field that has also influenced geoarchaeology, the landscape can be seen as "an assemblage of ecosystems –for example, forest, lakes, and streams – which also interact less directly than the components of ecosystems" (Russell 1997, 3). Interestingly, this definition not only suggests that the ecosystem is included in the landscape, but also that the landscape has a strong connection with the environment – the next contextual level (Figure 3.2).

The idea that the landscape is formed of natural and cultural elements, which is common in cultural geography and landscape archaeology, represents another issue for the interpretation of the record, particularly separating the natural and cultural features of landscape, which in many cases is difficult (Denham 2016; Wilkinson 2003). This situation complicates the already mentioned issue of determining the boundaries of archaeological sites. In fact, at the landscape contextual level site boundaries are diffuse or non-existent, particularly in agricultural landscapes (Chapter 11).

In terms of the geoarchaeological record, the landscape has visible and invisible features (Table 3.2), which represents another problem particularly in reconstructing the paleolandscape. The problem is that the paleolandscape is an abstract or virtual landscape because, unlike the modern landscape, it is reconstructed based on remnants of features dated to a particular period. The reconstructions sometimes extrapolate such remnants to create a more comprehensible dimension of an ancient landscape. For example, scattered remnants of paleosols are used as surface points to reconstruct the paleo-architecture of a flood plain.

The environment: the highest contextual level

The *environment*, like the landscape, is another term that can take several meanings. It can simply convey the general idea of the surroundings of a site; designate a depositional environment, which is referred to as "setting"; or it can have the meaning of an ecosystem. But in many cases the meanings are clear if they are preceded by a qualifier, or are implicit in the context of a narrative (e.g., social environment, physical environment, and fluvial environment).

For the purpose of characterizing the environment as a contextual level to interpret the geoarchaeological record, it is important to define it as all the abiotic and biotic aspects that interact with and integrate the other three levels (Figure 3.1a) and it serves also as the context of scale (Figure 3.1b). In this sense, the environment has an important property regarding scalability (Table 3.3), which can be adapted to different scales depending on the focus of research on the lower contextual levels (Figure 3.1b), thus serving as the basis for the idea of micro-, meso-, and macro-environment, used in general in the ecological context (*sensu* Butzer 1982). It is for this reason that the environment has an independent scalability, meaning that it is used as a means to focus on artifacts, assemblages, sites, groups of sites, settings, and landscapes, or even environments, depending on the scale of analysis.

Finally, like the case of landscape, the environment has a virtual part, the *paleoenvironment*, which with it is time-dependent, and can be interpreted as a dynamic term, that is, as an idea of environmental change in the past, or static as an idea of a particular environment surrounding an event or a situation fixed on a particular time period. This idea also implies that like the other contextual levels, the paleoenvironment can be reconstructed using any or all parts of the record (Table 3.2).

THE INTERPRETATION OF THE RECORD

Models and issues

After defining the geoarchaeological record, its parts, and the contextual levels at which it is studied, it is evident that the main issue to bring to discussion is the ways the record is interpreted. In reality, there is no particular way to interpret it, because it all depends on the the questions to be answered, the objectives of research, and the nature of the record itself. Thus, the topic of interpretation would require an entire volume.

However, it is perhaps important to briefly discuss the methodological approaches prevailing in geoarchaeological research: the stratigraphic and geomorphic approaches. But rather than explaining the basics of each of them, it is perhaps necessary to focus on how each of them addresses some of the interpretation issues of the geoarchaeological record and the particular aspects of each that may help discuss the topic in Chapter 4 (issues of interpretation).

To conclude this chapter, some thoughts on complexity and the use of technology are discussed, particularly with regard to sampling, correlation and interpretation. The objective of this discussion is not to provide a prescription to follow, but to recognize that technology sometimes has some limitations when untangling the complexities in the geoarchaeological record.

The stratigraphic interpretation of the record

Perhaps the most common model used for interpreting the record is the stratigraphic model, which has its variations depending on the practitioner's background and the object of study. Regardless, geoarchaeology looks at a stratigraphic record in terms of natural and cultural influences, particularly at the level of archaeological site. It is only in off-site contexts that the record is often seen from a geoscience point of view, but even so, it is almost always done at scales that encompass processes of relevance to human-environmental interactions.

The stratigraphic work in geoarchaeology also combines the study of soils, that is to say, it applies a pedostratigraphic approach (Holliday, Mandel, and Beach 2016). This approach acts differently at the various contextual levels. At the level it focuses on aspects of soil formation that affect the archaeological record. At the level of setting, pedo-sedimentary sequences are often useful in establishing sequences of events, if not also dating. At the level of landscape, the pedo-stratigraphic approach focuses on landscape stability and paleo-landscape reconstruction. At the level of environment, it provides a proxy for reconstructing a series of environmental factors (e.g., climate).

The influences of archaeological stratigraphy in the analysis of stratigraphic sequences in geoarchaeology are evident, despite the fact that many practitioners are unaware of them. Among the many influences is the idea of seriation in the presence of cultural deposits, which forces the geoscience-based practitioners to look at cultural features, and expect in many cases deposits that, although looking natural, are in fact artificial (see Chapter 4).

The Harris Matrix, despite its criticism (see Goldberg and MacPhail 2006), has also influenced the way the stratigraphy geoarchaeological record is studied. The section shown in Figure 3.6, although it represents a geoarchaeological setting, has used information from 14 other sections in the area to create a matrix at a particular type locality. The matrix can perfectly be applied to non-cultural stratigraphic settings, where erosion and depositional hiatuses are also common.

The influence of geomorphology: The ten ways to be wrong

Geoarchaeologists in the geomorphological and soils tradition have turned to geomorphological models developed for explaining the cause-effect process of landscape development and site formation (Frederick 2001; Holliday 2004; Borejsza and Joyce 2017). One of the several causative models that geomorphologists followed in recent decades is the causative model known as "Ten ways to be wrong," devised by Stanley

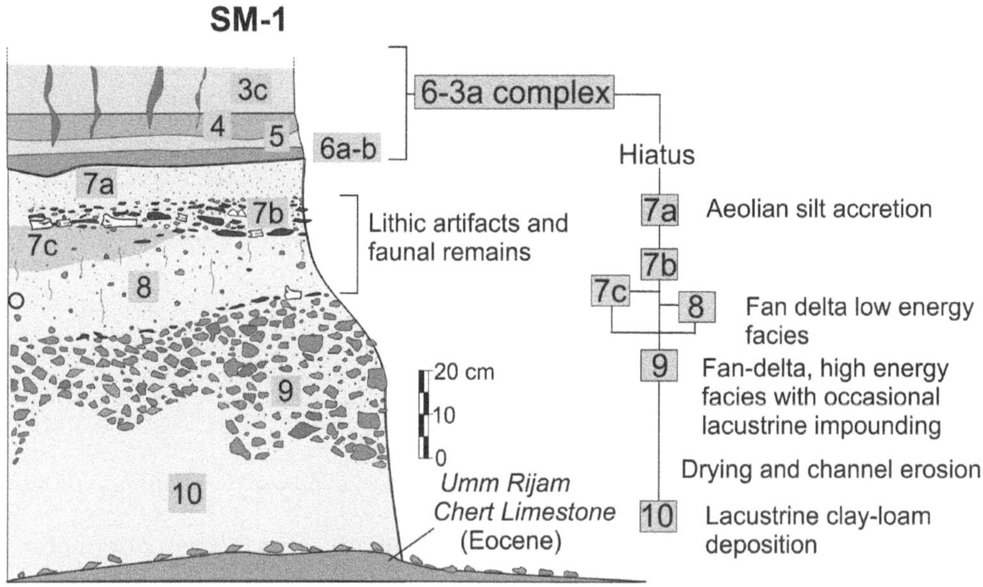

Figure 3.6. Relationship between sedimentary units and time-seriated geomorphic/depositional events associated with a late Middle Pleistocene, archaeological layer in Azraq-Shishan, Jordan (From Nowell et al. 2016, with modifications).

A. Schumm (1991). The model focuses on ten persistent problems of interpretation of the geomorphological record and ways to overcome them (Table 3.4). The ten problems can be grouped into three classes depending on the broader sets of issues faced: scale and location, causes and processes (i.e., causality), and the system's response. Although geomorphological in nature, the problems have been adapted to the interpretation geoarchaeological record (Borejsza and Joyce 2017; Contreras 2017), with some interesting examples discussed in Chapter 4.

Most of the issues in geoarchaeological interpretations seem to be related to how we see *time* in relation to *space* and the scale of time we use. This is an issue particularly between geology and archaeology, where time is often measured at different scales. But beyond this difference, the background disciplines and fields sometimes vary in terms of where they put time in relation to space. For example, in geology time is a more important factor than it is, for example, in geography, which sees space take the lead as a primary factor.

As pointed out by Alistair F. Pitty (1982) the concept of time plays different levels of importance in the different disciplines associated with geomorphology (Table 3.5). This idea can be transposed to the multidisciplinary conglomerate that forms geoarchaeology. Thus, time might be of more interest of a practitioner working on archaeometry, as opposed to one trying to solve a problem of spatial distribution of features in a landscape.

Time-related issues include, first, problems of period and span (Table 3.4), with *period* being a pre-defined time frame, and *span* the time it takes for an event to occur. Space, on the other hand, represents other issues with scale and size, with *scale* meaning the degree of focus with which we see a phenomenon, and *size* meaning the magnitude of the

Table 3.4. Ten problems in geomorphology divided by classes and with their variants (Schumm 1991)

Class	Problems	Variants
I Problems of scale and place	Time	Period and span
	Space	Scale and size
	Location	Place and perspective
II Problems of cause and process (i.e., causality)	Convergence (equifinality)	Cause and process
	Divergence	Cause and process
	Efficiency	–
III Problems of system response	Multiplicity	–
	Singularity	–
	Sensitivity	–
	Complexity	–

Table 3.5. The priority of time among the dimensions across different disciplines associated with geomorphology (Pitty 1982)

Earth Sciences				Distributional Sciences	
Geology	**Hydrology**	**Pedology**	**Geomorphology**	**Geography**	**Statistics**
time	vertical	vertical	vertical	length	vertical
vertical	*time*	width	length	width	width
width	length	length	width	vertical	*time*
length	width	*time*	*time*	*time*	length

phenomenon. Location refers to aspects of place from the researcher's perspective based on the position of the object or phenomenon in a coordinate and altitudinal system.

The combination of the concepts of time, space, and location are very important when interpreting the geoarchaeological record, because of the types of issues one can encounter. For example, the occurrence of time-transgressive phenomena means that event has a lag between *time, space,* and *location,* thus creating a perceived delay in the effect or the response of phenomenon.

Despite being proposed for geomorphology, Schumm's ten problems constitute a conceptual basis for the interpretation of the geoarchaeological record, even in cases that do not involve geomorphological processes. For example, the problem of *convergence* (equifinality) can be extended to the discerning N (natural) and C (cultural) processes are often one aspect that geoarchaeologists have to interpret, a matter that has been often discussed (e.g., Butzer 1982; Stein 2001a; Rapp, Jr. and Hill 2006). But its antithesis, divergence, is also important since convergent results cannot be explained if there are no divergent results.

Problems of *singularity, multiplicity,* and *complexity* are very important for understanding site formation processes, because such processes are usually complex and are the result of multiple causes. These three problems are practically the norm in most archaeological and geoarchaeological contexts at different scales.

Technology and the interpretation of the record

In recent years, however, the availability of technical aids has provided geoarchaeology with tools to solve problems of complexity, among others, as well as creating models that would serve as means to bridge the gaps posed by missing data (i.e., absent parts of the record).

The use of technology, although a great help in the interpretation of the record, can lead to misinterpretations. One of the issues with GIS in environmental archaeology is when scales of analysis are shifted without considering resolution, a practice that can lead to erroneous conclusions and fallacies (Harris 2006; Zimmerman and Artz 2006).

Sometimes low spatial resolution is the result of missing data or poor or sparse sampling. which is an aspect that needs to be considered in any spatial analysis. Therefore, judgement of data resolution problems is an aspect that transcends technology, which in many cases is part of the process of improving methods of data collection.

4

The Geoarchaeological Record

Interpretation Issues

VISUALIZING TIME, CAUSALITY, AND CONTEXT

Causality issues in the geoarchaeological record

Geoarchaeology is a historical discipline in the sense that it aims at reconstructing the past, a characteristic shared by its parent disciplines, archaeology and geology. In both these fields, the reconstruction of the past can be *static*, that is to say, representing the relation of events at some point in time, or *dynamic*, involving processes and events occurring at different scales, a view that involves cause-effect relationships between processes and events. But although it sounds simple in theory, the interpretation of causality in the record may face several issues inherent to processes of formation of the record itself.

One such problem is the fragmentation of the record, which poses problems with connecting events. A second problem is the degree of complexity that cause-effect processes entail in terms of convergence, divergence, complexity, and multiplicity (Table 3.4). A third problem, and one that is inherent to the nature of the geoarchaeological record, is the intertwining of natural and human phenomena, which also become entangled with issues of convergence and divergence, making difficult to determine the natural or human causes of events in the record. Regarding the three issues outlined above, three concepts become the focus of understanding causality: *process, event,* and *spacetime*.

The concept of process is better understood because it is less ambiguous than event and spacetime. The main aspect to consider when studying a process is that it is often viewed as an isolated phenomenon. Thus, a fluvial process produces a fluvial landform. But it is not as easy as it sounds for the same problems discussed above, that is to say, complexity, convergence, and divergence, among others. In reality, landforms are created by a series of processes, even if they are considered just fluvial, glacial, or eolian. The matter becomes more complicated as geoarchaeology deals with geomorphic processes in parallel with site formation processes, and even social and environmental processes at various scales. For this reason, processes should be viewed in the context of events and spacetime.

Events in space and time

The term event has different nuances in the different disciplines. In geology, it is often events are seen as momentary changes such as the onset or cessation of a process (Engelhardt and Zimmerman 1982), which conveys the idea of rapid change. In geomorphology, the idea of event is linked to process, which is often seen as an observable dynamic phenomenon (Brunsden 1996). In archaeology and history often events are seen at human scale, which represents sometimes minute fractions of geological events (Stein and Linse 1993).

In a more general view of time, an *event* is defined as "a happening process, typically extended over both time and space" (Dainton 2010, 431). Sometimes the terms *event* and *episode* are used interchangeably, and many dictionaries say the two terms are synonyms. But usage in the literature distinguishes them clearly. An event is an occurrence, while an episode can be merely an incident within a longer occurrence, as for example the case of short period of calm (an episode) within a storm (event). In the geosciences, for example, an episodic occurrence is an event with several small rapid events (Brunsden 1997).

An event can also have different connotations depending on whether it is seen as a time span or period (Table 3.4). *Span* is the time that takes for a process or an event to occur, while *period* is the time frame that we subjectively assign to studying a process or an event. The relation between time and span can be seen for example in the case of the Younger Dryas (YD), a cooling event that is recorded in the pollen records and glacial records of Europe and North America and in the ice core records of Greenland. Accordingly, the YD has been conventionally placed in a time frame between ca. cal 12,900 and 11,600 years BP – a period. But this time frame does not necessarily match the time span of its effects across the globe. This example also shows that concept of period ties an event to time, while span ties the event to both time and space. In other words, the period is meant to simplify problems created by lag effects across space.

In the philosophy of time, the relation between objects or events within time is represented in two ways, the *substantivalist* way in which objects or events navigate through a medium tied to a timeline, and the *relationist* way, in which the objects and events are tied not to a timeline but to each other (Dainton 2010; Meyer 2013). In both cases, however, the idea of time implies a four-dimensional way of thinking, referred to in the literature of time mathematics as *spacetime* (Rucker 1984).

Examples of event relations in geoarchaeology are often represented as abstract diagrams showing the development of events through times. One classical example that approaches substantivalist form of space in geoarchaeology can be represented by a stream valley changing through time, where each section indicates an event (e.g., incision and deposition) or as the seriation of stream valley events as part of a series of other events (Figure 4.1). It is only an approximation to the substantivalist idea of time, because in theory the real representation would include thousands, if not millions, of sections to show the changes in the stream valley. The process is then abstracted to just a few sequential images. Other examples that approach the substantivalist representation of time are series of settlement maps by archaeological period. Like the case of the stream valley sections, each map is only an abstraction of objects or events for each period.

Examples of the relationist representation of time in archaeology and geoarchaeology exist in the form of chronological diagrams, which show the simultaneous process of different events such as climatic trends, cultural changes, and geomorphic processes.

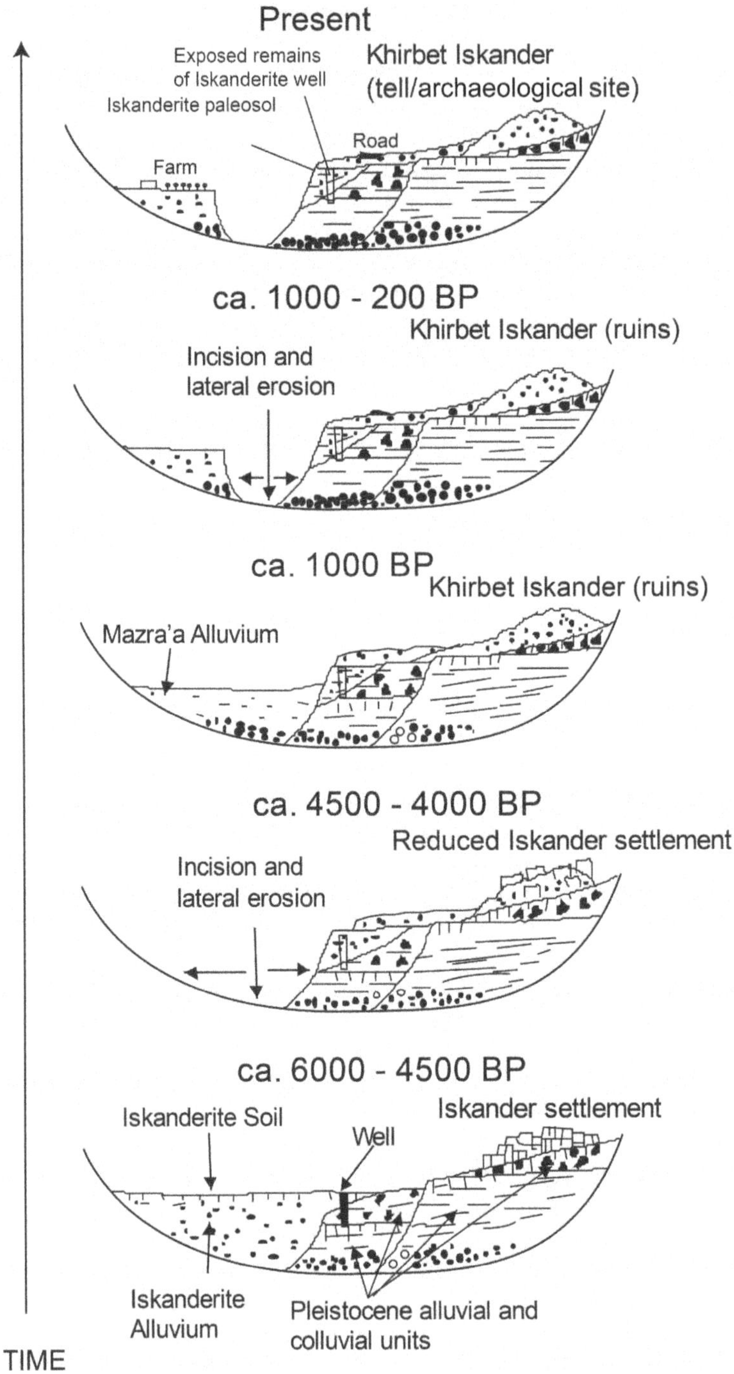

Figure 4.1. Incision and fill events in the floodplain of Wadi al-Wala, adjactent to the Early Bronze Age site of Khirbet Iskander, Jordan. Modified from Cordova (2007, 2008).

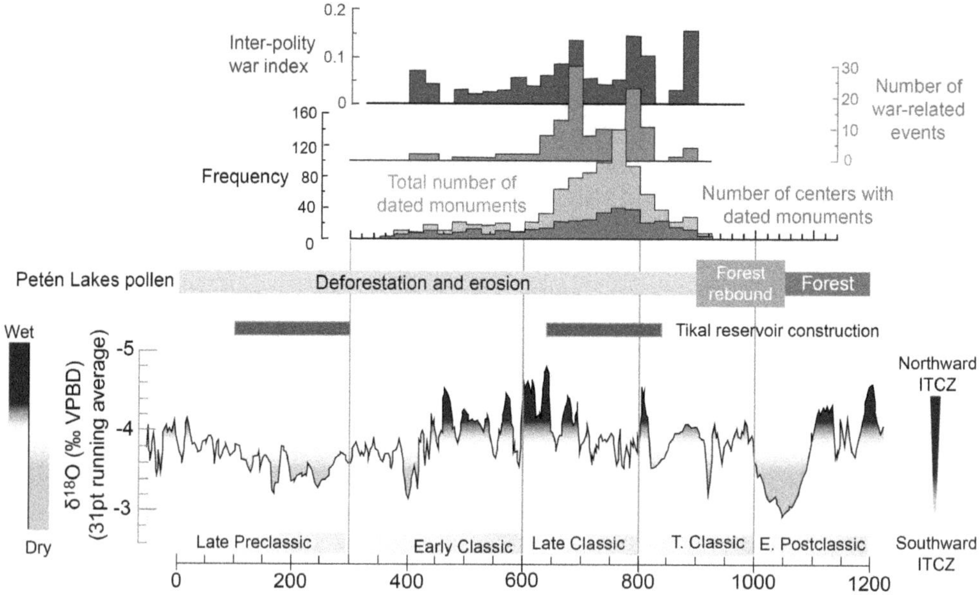

Figure 4.2. Cultural events in the Maya Region in relation to environmental changes and precipitation fluctuation. Composite is after Kennett and Beach 2013, and data from various authors (Anselmetti et al. 2007; Kennett et al. 2012; Scarborough et al. 2012).

The events in the time sequence depicted in Figure 4.2 are meant to be related, thus constituting and example of the relationist view of time. Even if the ages are removed from the bottom diagram, the events in each of the graphs would bear a correlative relationship. Thus, the effects of precipitation would still maintain a causal relation with the cultural developments in the Maya region from the Preclassic to the Postclassic (see Case 10.2, Chapter 10). The example also shows that the relationist conception of time is useful as means of relating events to explain other events, which also works in terms of establishing causal relations between events and processes.

Sometimes separate events are produced by different causes can act together as catalysts for other events. Take the case of the Dust Bowl shows that the 1930s, for which the drought that triggered the event was totally independent from events that started at the beginning of the century (e.g., plowing of grass sod) and the Great Depression of the 1930s. Nonetheless, the three events converged to create the Dust Bowl crisis (Case 14.2, Chapter 14). Similarly, the devastation of New Orleans by Hurricane Katrina in 2005 was not only the result of the hurricane itself (i.e., the main event) but a series of events unrelated to the hurricane came together to cause the catastrophe (Case 13.2, Chapter 13).

The spacetime relationship in history and archaeology has been discussed in its relation with time and space and with other events, but many times we do not question what the idea of *event* encompasses across disciplines and scales. As mentioned in the geological and geomorphological definitions of event above, it is clear that event and process in most cases can be synonymous. But this is not the case in the social sciences and humanities, and particularly in history where events have different nuances, marked by scale, but interestingly different scales as those in the geosciences.

French historian Fernand Braudel (1980) separates events into three categories. The first category is *longue durée* (the long term), which denotes the idea of continuous process through a long period of time, often referring to long-term events that continue into the present, but often too long to be perceived by human individual because they exceed the human's lifetime. The second category is the *evenement*, which translates as "event," and designates a process that begins and end normally in a shorter time, normally perceptible in a human's life time. The third category is *conjuncture*, which conveys the idea of a convergence of events that creates a new event. Historically, for example, one can see capitalism as a *longue durée* process, and World War I as an event and a conjuncture of events that in some countries and regions of the world created a new order that somehow persisted through the rest of the century.

The Braudelian view of events, although appropriate in terms of describing events in the history of civilizations, fails to provide units of time for processes at meso- and micro-scales, which are common in archaeology and geoarchaeology. In later writings, Fernand Braudel introduced and used the idea of levels of planes (geographic, social, and individual levels) instead of the three categories of events explained above. The geographic, social, and individual levels (Orser, Jr. 2014). The levels, however, are not necessarily equivalent with the types of events above, but do portray the idea of time scale in a better manner. The problem is that the so-called geographic level, the one that is supposed to include the environmental changes, is seen as the less perceived by the individual. This is not necessarily true because there are situations that may be remembered depending on the intensity of phenomena, as is the case of individuals remembering a natural catastrophe, but not other events.

In geology, the idea of reconstruction of events out of the stratigraphic records seems to be more widely accepted, if not taken for granted. In archaeology, on the other hand, the interpretation of stratigraphic records often draws more debate on aspects of perspective and chronological seriation. One the one hand there is criticism of the idea of events as discrete or punctuated happenings in time, and, on the other, the problem of perspective.

The punctuated idea of events has led to the problem of *cumulative palimpsests* (Lucas 2008), which are not necessarily the same as the stratigraphic palimpsest discussed below in this chapter, but to the practice of accumulating events in a linear form, a practice that was typical of the old archaeologies (e.g., pre-1950s). Nonetheless, this idea is still latent in modern interpretations of time in archaeology.

The problem of perspective is called *time perspectivism* (Bailey 1983), which sees and describes the occurrence of events in time periods and not in time spans, representing a potential problem of misinterpretation of the sequence of events in the archaeological record. One of the problems is the so-called *Pompeii premise*, which in broad terms is the idea that archaeologists tend to base their interpretations of their finds as if they were frozen in time, leading to diachronic reconstructions of events. Time perspectivism is not unique to archaeological interpretations, it is a problem seen sometimes in geoarchaeological interpretations, particularly when natural and cultural phenomena are boldly matched. The case of the Xitle Eruption and its effects on the Preclassic settlements of the southern Basin of Mexico is an example that represents this situation (Case 13.1; Chapter 13).

Needless to say, the reconstruction of events in space and time faces issues of complexity, which are inherent to the cause-effect nature of multiple events happening at the same time (Figure 4.2). If not clearly defined in the record, these complexity issues can lead to chicken-or-egg situations, as often occur in the study of climatic and human

Figure 4.3. (a) Laminated sediments exposed on a road-cut section; (b) close up to show the horizontal lamination; and (c) remains of ancient dam.

influences in certain processes such as soil erosion, salinization, and many other phenomena that appear in the geoarchaeological record.

Causality problems need to be solved at all contextual levels, from those of site formation processes to those related to global environmental change. The following sections focus on some issues that are commonly encountered in the record, for which there is in many cases no definite solution, but at least should be taken into careful consideration when interpreting the geoarchaeological record.

CAUSALITY IN NATURAL AND CULTURAL TRANSFORM PROCESSES

The definition of natural (N) and cultural (C) transform processes by Michael Schiffer (1987) is clear when their effects are considered, but difficult when it comes to establishing their causes. In part this problem lies in the difficulties in establishing what is purely natural and what is purely cultural, merely because there are several types of processes within each category.

The cultural processes have two kinds: those influenced by behavior, which are the primary interest to the archaeologists, and those that alter or obscure the original behaviors, while often non-cultural are environmentally produced actions that alter, obscure, or preserve the original behavioral signatures in archaeological deposits (Stein 2001b). However, cultural processes can alter natural depositional systems, which in turn may create deposits that, despite being artificial, look natural when seen out of their geomorphic and cultural contexts.

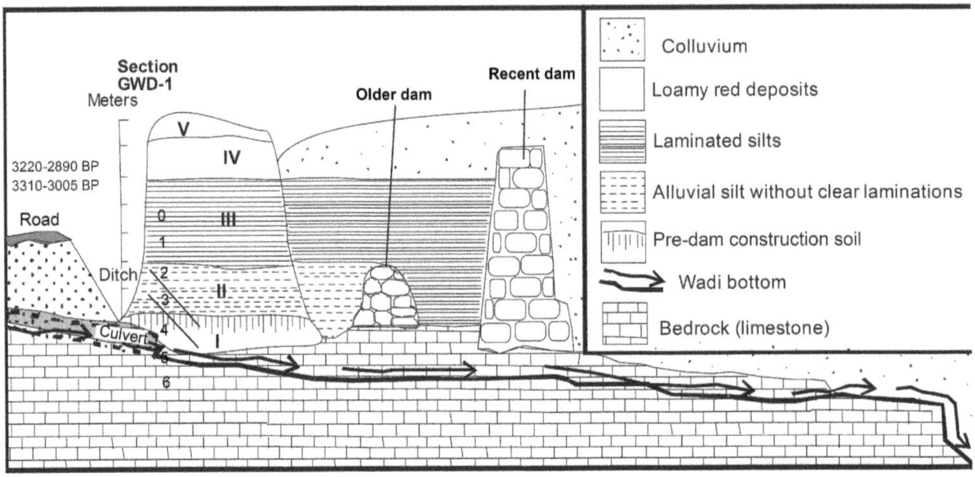

Figure 4.4. The contextual setting of the Dhiban dam and its sediments.

By appearance, the deposit in Figure 4.3 looks like a laminated lacustrine facies (Figure 4.3a–b) but, when put in the context of the local stream and the remains of a dam structure in the vicinity (Figure 4.3c and 4.4), it is clear that it is the deposit of sediments trapped in a reservoir. Research of the record at the setting and landscape contextual levels revealed that the dam was part of a series of dams established across Wadi Dhiban, on the Western Jordanian Plateau, by the Nabatean settlers of the nearby Tell Dhiban (Cordova 2010). The example shows that indirectly the deposits were created by low-energy alluvial accumulation (N-transform process) after the manipulation of the stream by building the dam (C-transform process).

There are cases where accidental natural deposits are created by management of structures may or may not be considered natural transformation processes. For example, the floods of 1993 in the Upper Mississippi were caused mainly by the raising of levees upstream, which increased the flow of the river by not allowing water to spill over backswamps, now built up or farmed. Then the breaking of the artificial levees downstream caused immense flooding and deposition of silts. If a geoarchaeologists with no historical background of the 1993 floods look at the deposits engulfing house and field remains, he or she will interpret the deposits as an overbank silt deposit caused by a natural flood, not by a flood directed to the location by the construction of an artificial levee, technically a C transform processes.

One problem in establishing causal relations from the record occurs when the anthropogenic influence in the causation of natural phenomena is often unrecognized in cases when soils and sediments are devoid of artifacts. Only a contextual approach, often involving multi-scale views, can clarify causal relationships of phenomena interpreted from the geoarchaeological record.

TIME-TRANSGRESSIVE PHENOMENA IN THE RECORD

Time-transgressive phenomena have received attention in geological stratigraphy, particularly in places where the limits of certain lithostratigraphic units across space and

Figure 4.5. Tesesquite Creek, Cimarron County, Oklahoma; late 1960s and 2004 (Wilson 1972; Cordova and Porter 2015).

time. The evolution of the landscape, in the geomorphological sense, can also be affected by time-transgressive phenomena as geomorphic systems vary through time.

Very common cases of time transgressive phenomena occur in the response of fluvial systems to adjustments, as is the case of channel entrenchment, which often occur at different times along the valley. The case represented in Figure 4.5 shows the historical development of a typical arroyo entrenchment, initiated downstream in the mid-twentieth century and reaching its higher reaches several decades later (Cordova and Porter 2015).

The channel trenching event in Tesesquite Creek is not only time-transgressive within the stream itself, but also in the broader regional context. Channel trenching in the broader American Southwest occurred mainly towards the end of nineteenth century (Cooke and Reeves 1976; Vogt 2016), which makes Tesesquite Creek, located in far-western Oklahoma, out of synchrony with the regional process. The interesting aspect of it is that the process was initiated in the nineteenth century and has affected the broader southwest regions ever since at different times. Yet, synchronicity between basins is sometimes not the case, as shown by some cases in southern Arizona (Waters 2008).

This situation has been a problem in the sand dune stratigraphy and geomorphology of the Great Plains, where the trend has been to correlate sand dune development with broad regional phenomena, when sand dune development responds to more local phenomena (Halfen and Johnson 2013). This is the case of the actualistic example of the 1930s Dust Bowl, an event that seems inexistent in the stratigraphic and geomorphic record seems inexistent of many areas where it occurred (Case 14.2, Chapter 14). The problem is that when trying to correlate events, one forgets that causality and scale are more important than connecting events to causes across scales (Contreras 2017).

The geoarchaeological record also shows cases of time-transgressive cultural processes. The case of the Late Epipaleolithic of the Levant, the precursor of the first Neolithic communities in the Southern Levant, presents several forms of sedentism that predates plant and animal domestication (Case 8.1, Chapter 8), which interestingly is a complex process where some aspects of behavior are intertwined with local responses to changes in climate and landscape, which interestingly differ from other areas within the same region

(Maher 2010; Jones et al. 2017). Thus, it is important to bear in mind that time-transgressive phenomena do occur when looking at cultural processes and certainly in the development of human-environmental processes across a region.

ARCHAEOLOGICAL VISIBILITY, INVISIBILITY, AND ABSENCE

The reconstruction of sites, activities, paleosurfaces, processes, events, paleolandscapes, and paleoenvironments is hampered by the fragmentation of the record, which, in concrete terms, refers to its invisible and absent parts (Table 3.2). Invisibility and absence are often confused and taken for one another, a common case that occurs particularly at the regional level of landscape, as illustrated in the following example.

Archaeological survey maps of the Basin of Mexico have been an important source of information compiled over several decades (Sanders, Santley, and Parsons 1979) and made available for the interpretation of settlement patterns. The survey maps show extensive areas in the plains around the lakes that had no Prehispanic sites or surface remains at all (see a sample in Figure 4.7). Various a priori explanations tried to explain the "absence" of sites, some suggesting that these areas were too risky for agriculture, too difficult to farm because of heavy soils, or areas that at the time may have been inundated by the lakes.

In the decades following the surveys, the digging of sand pits and brickyard pits exposed several sites with floors buried under several meters of alluvium, most of which was deposited in the seventeenth and eighteenth centuries as revealed by geoarchaeological research in the Texcoco and Chalco regions in the east and southeast of the Basin of Mexico (Hodge, Cordova and Frederick 1997; Cordova 2017). In Texcoco, areas with no prehistoric archaeological remains on the surface coincide with areas of floodplains where overbank and crevasse-splay deposition has been intense in historic times (Figure 3.4d and 3.5).

A similar situation appeared in the Chalco region along the former flood plains of the Tlamlmanalco and Amecameca Rivers and the delta of the former (Figure 4.6a). At the southern end of the delta, the finding of late Aztec lacustrine raised fields (*chinampa*) (Figure 4.6b) compared with historical maps suggested that the delta prograded rapidly during the fifteenth and sixteenth centuries. These findings, which would have not been possible without *off-site* geoarchaeology, exposed the invisible parts of the record and complemented the reconstruction of the Prehispanic landscape of the alluvial plains around the lakes of the Basin of Mexico.

Unlike the apparent concealment that invisibility entails, absence means something has been removed from the record. However, in some cases *traces* in the record may suggest the previous presence of a feature. For example, in the southern Levant valleys there is hardly any material evidence (dams, canals, and other structures) of irrigation in early agricultural times (Neolithic and Chalcolithic, ca. 10,000–5500 BP) and even early urban times (Early Bronze Age ca. 5500–4000 BP) (Rosen 2007a; Cordova 2007). Nevertheless, traces in the record suggest that irrigation must have been practiced in the flood plains adjacent to settlements, despite the lack of material evidence in the form of remains of canals and dams. Indirect evidence exists in the form of cereal phytoliths, which, based on experiments, suggests that wheat was cultivated under irrigation. In this case, the absence of irrigation installations associated with early agricultural periods can only mean that the reconstruction of irrigation can be achieved through traces, as mentioned above, or through hypotheses based on (a) modern traditional irrigation, or (b) on the observation of fluvial processes in wadis today, which are perhaps responsible for deterioration (Cordova 2007).

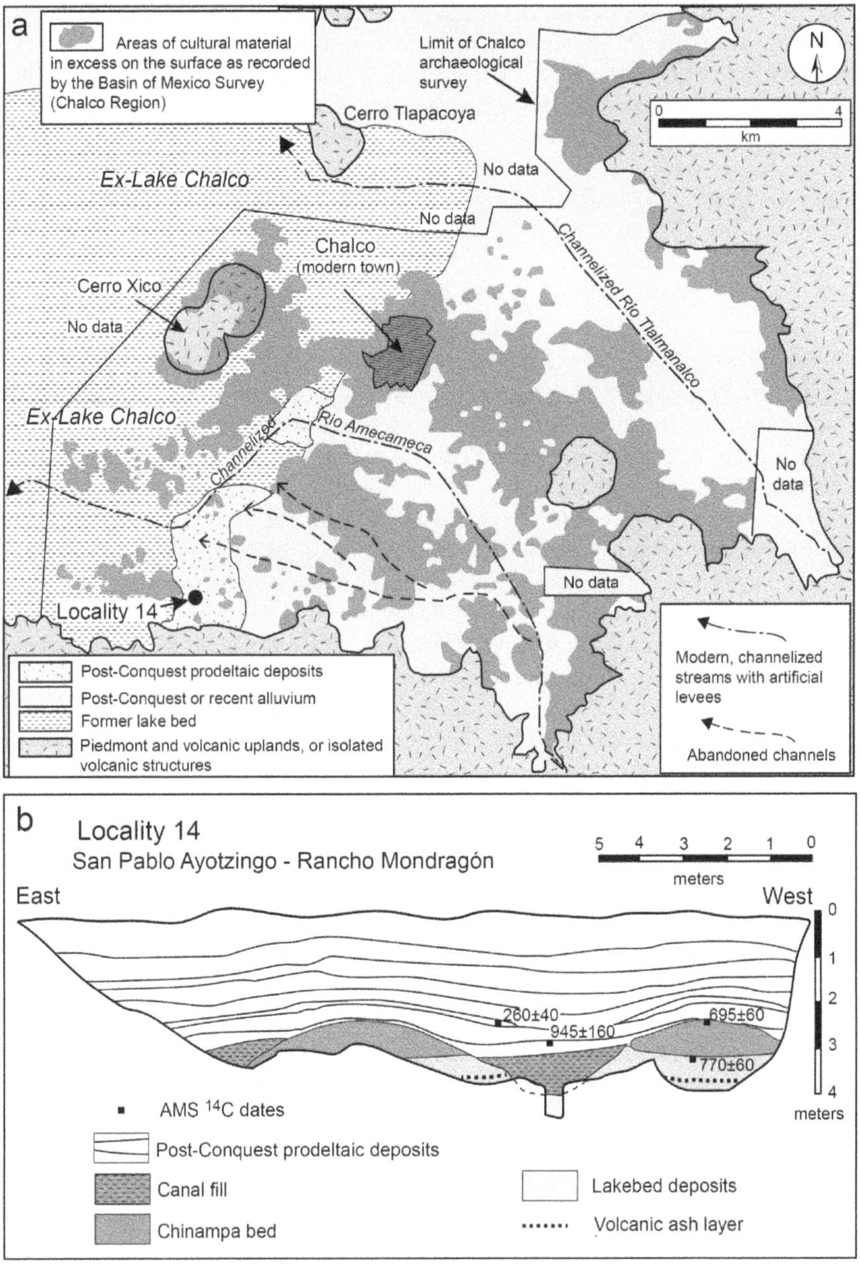

Figure 4.6. a) Map of the Chalco Region, Basin of Mexico, with survey areas with no-site areas (From Hodge, Cordova, and Frederick 1997); b) buried Chinampa near Ayotzingo (see Locality 14 on map). The chinampa, built on a shallow lakebed in the Late Aztec period, was buried by the historic progradation of the Amecameca River delta (From Hodge, Cordova, and Frederick 1997).

Finally, the invisible and absent parts of the record are sometimes reconstructed through hypotheses and very often through *abduction* (discussed in Chapter 2), both of which entail a reconstructive process that today is done through careful recording of the visible parts, the exposed non-visible parts, and the traces of past objects and processes. Although this is now often done by way of computer simulations using state of the art technology it is important to understand that such models are hypothetical, particularly when the absent evidence is sizeable.

THE VIRTUES OF OFF-SITE GEOARCHAEOLOGY

The objectives of off-site research in geoarchaeology are straightforward – they encompass the compilation of information for the reconstruction of paleolandscapes and paleoenvironments. This information is often not available on-site for a number of reasons, of which the most important is the noise created by human activities on environmental processes.

Although the advantages of off-site geoarchaeological research are clear, some practical problems need to be acknowledged. The most common of such problems is the lack of sediments and soils that could provide information for paleoenvironmental reconstruction. Typical cases occur in areas of highly dynamic geomorphic environments that prevent the deposition of sediments and the formation of soils. But other biological and geomorphic situations can also prevent the formation of a sedimentary record.

Another practical problem occurs in places where sediment accumulation is considerable, but there are not exposures to study the pedo-sedimentary record. Although non-destructive techniques for site prospection underground exist (e.g., magnetometry, resistivity, and ground-penetrating radar), they are not good for the study of sedimentary sequences, which require sampling for age determination and other proxies, and thus require more invasive techniques such as coring or digging.

Coring is possible only through soft and fine sediments, and multiple coring can provide information for reconstructing paleosurfaces and correlating strata. The problem is that coring would not permit direct visual contact with nuances in the stratigraphy, particularly small horizontal features and short-distance variations in layers. Coring can also be used as a prospective tool to assess the local stratigraphy of a site or around sites.

The practice of backhoe trenching, widely common particularly in CRM geoarchaeology, is often applied in areas of thick sediment accumulations where digging by hand (i.e., test pits) would require significant time and labor (Ferring 2001; Holliday 2004). However, from the point of view of archaeology, backhoe trenching is invasive and destructive, and its practice is sometimes not welcome by some archaeologists because of the possible obliteration of archaeological information. Yet, despite this problem, backhoe trenching can reveal the existence of sites in areas where previously no sites existed, as in the example discussed for Figure 4.6.

Because of the destructive nature of backhoe trenching, decisions on where to core or dig without destroying sites need to be assessed by the team. A strategy to avoid destruction is to be vigilant while the trenching is done, and look for any evidence of floors or walls as digging proceeds. If a structure is hit, then backhoe operations can be called off and careful digging ensue.

Off-site geoarchaeology may not be possible in some situations, particularly in CRM work limited to a particular locale, or in cases when budgetary constraints prevent the expansion of research into the off-site realm or if off-site information is not considered a priority. But where it can be done, off-site research must be done, because it often provides relevant information about the context of occupation.

LEGACY EFFECTS, RELICTS, AND PALIMPSESTS

Legacy effects of relict and human features in the landscape

The term *legacy* implies the idea that something left from the past influences what happens in the present. For example, a landform left from past erosional processes (e.g., a paleosurface) may determine the processes that act today at that particular location. Likewise, glacial sand or loess deposits laid in the past may influence the way the landscape has subsequently evolved to the present and will into the future. The term *relict landform*, for example, implies the idea that the landform was created by processes in the past that no longer act, but the term does not necessarily convey the idea of influence in the present. However, a relict landform or a cultural feature can become a legacy if the landform is used by subsequent settlers as an advantageous point in the landscape or if a relict feature is re-used by generations of settlers.

Although sparsely used in the geoarchaeological discourse, *legacy* and *relict* are adjectives that qualify landforms and sediments with potential use in the description of settings, landscapes, and processes of change. Unfortunately, when used in front of terms such as sediment they take a different meaning that the geoscience literature has in fields totally different from geoarchaeology.

The meaning of *legacy sediment*, which would in general mean a sediment left from the past that influences subsequent processes, has in recent decades shifted to refer to designated an anthropogenic sediment, particularly in studies of pollution in streams (James 2013). The problem with this meaning is that that natural processes also leave legacies. Similarly, the term *relict sediment*, although with a general meaning in its beginning, has been applied by geologists to designate terrestrial sediments of the continental shelf drowned by the post-glacial sea level (Davis, Jr. 1992). The problem here is that other geologic processes create relict sediments. But beyond the overlaps with meanings in other geoscience fields, geoarchaeology can reclaim such terms for a proper use with ampler meanings.

There are several cases of legacies and relicts in almost any landscape. Geomorphic and sedimentary legacies in the Great Plains of North America include, for example, the large amounts of alluvial sands deposited in the Pleistocene, some of which have been reworked by wind into dunes several times in the Holocene (Muhs and Holliday 1995). Some stabilized dunes, as is the case of the Nebraskan sand hills, are both legacies and relicts of the glacial past. Some of these landscapes created important settings for particular types of occupations in localities where geomorphic and hydrologic processes were combined to attract fauna and humans, as is the case of the Blackwater Draw locality, near Clovis, New Mexico (Holliday 1997; Haynes 2005).

Tectonics in combination with sea level change can also play an important role in leaving relict landforms, such as escarpments, valleys, and a number of features that may play a role in subsequent events shaping the natural and cultural landscape. Paleovalleys in southwestern Crimea, for example, were not only important for the ancient Greek farmers who settled there, but also as sediment traps that contain proxies for paleoenvironmental reconstruction (see Case 11.1, Chapter 11).

Relict soils can also be important in the geoarchaeological record, as the properties inherited from older pedogenic developments become available in more recent times (Birkeland 1999). Such properties may play an important role in shaping more recent landscapes or in the development of agriculture of recent ages. In the western Jordanian Plateau, for example, the Pleistocene red soils are paleosols that have been exhumed by deflation and used by modern and most likely by past rainfed agriculture due to their texture and capacity to retain water (Cordova Cordova et al. 2011). The red soils were formed under conditions remarkably different during the MIS 5 and early MIS 4 under conditions remarkably different from those of recent times (Cordova et al. 2011).

Identifying legacies and relicts and their role in the evolution of the landscape is important when studying the geoarchaeological record, particularly by identifying properties that may cause attraction to human use. Legacies, as suggested by the examples above, are important in determining certain problems with dating processes and events. Again, legacies and relicts of the past exist in almost any landscape on earth.

Palimpsests and their issues of interpretation

The concept of a *palimpsest* in stratigraphy draws on the ancient practice of writing, erasing, and rewriting on several materials such as wax tablets and manuscripts, in which case fading traces of the old writing remain. In the archaeological record part of what was meant to be erased still exists as residuals (Lucas 2010). One can think of a palimpsest as an accumulation of artifacts from several ages in a thin layer, where the sedimentary matrix has accumulated slowly over time, or where the sediment has been constantly removed (i.e., the idea of erasing). Seen in the stratigraphy, a palimpsest is a thin accumulation of sediment encompassing several ages. In this sense, palimpsests in stratigraphy are a middle point between completeness and hiatuses, or between unconformities and conformities (see, for example, Figure 3.6).

Although the concept of palimpsest is largely applied to the archaeological stratigraphy, palimpsests are evident in archaeological landscapes (Wilkinson 2003). In fact, landscape is generally a palimpsest, where features of various ages are horizontally distributed. Such features in the landscape are used, abandoned, re-built, and re-used, sometimes belonging not to one, but to several, archaeological periods.

In view of the above, one can then distinguish at least three types of palimpsests in the record: a) cumulative palimpsests; b) erosional palimpsests; and c) landscape palimpsests. The cumulative and erosional palimpsests can be envisioned in an idealized diagram with the interplay of four variables: rate of sediment accumulation, sediment erosion, rate of pedogenic development, and artifact accumulation.

If the rate of sediment accumulation declines at the expense of the erosion rate while artifact accumulation and pedogenic development remain constant (Figure 4.7a), a sequence of layering or clustering of artifacts will clearly be seen (Figure 4.7b), in the context of a pedosedimentary model (Figure 4.7c). In the latter two (Figure 4.7b and c), numbers 2 and 3 could be considered cumulative palimpsests, while numbers 4 and 5, erosional palimpsests.

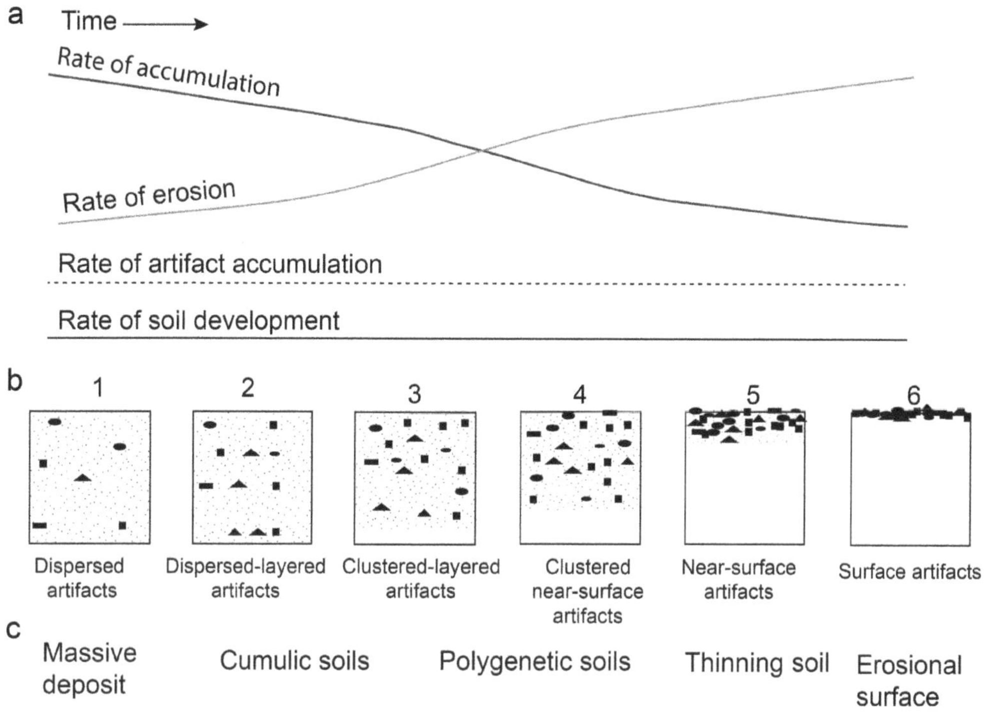

Figure 4.7. (a) Artifacts in relation to cumulative and erosional palimpsests, where the accumulation of artifacts and rate of pedogenesis remain constant; (b) schematic expression in pedo-sedimentary sequences with artifacts; and (c) typical characterization of the pedo-stratigraphic unit. The cumulative-erosional processes are assumed to be either water or wind action.

A cumulative palimpsest is more like the case of its meaning in archaeology, where natural and human residues accumulate (Terry 2016), then providing a relatively thin layer with sediments and artifacts of different ages. An erosional palimpsest would constitute a mix of artifacts of several ages.

In natural contexts, however, palimpsests can develop differently, particularly when accumulation rates are low and soils develop on stable landscapes where polygenetic soils are common (Holliday 2004). Polygenetic soils would be at the point where the rate of accumulation is low, forming a spectrum from sediment, through cumulic soils, to eroded soils (Figure 4.7c). In geoarchaeology, the important aspect to recognize in soils is that if, artifacts are accumulated in a polygenetic soil, then the soil would also be an archaeological palimpsest.

A palimpsest in the landscape, as described by T.J. Wilkinson (2003), can occur as "different levels of preservation and loss of individual features through time have resulted in any given landscape comprising a wide range of features dating from different periods. [The] fundamental notion of landscape as palimpsest deals with the progressive superimposition of one landscape on another and sometimes the selective removal of part of the earlier landscapes by later landscapes" (Wilkinson 2003, 7). This statement makes the point of addition, re-use, and removal of features in the landscape an idea that also has an important connection with the problem of absence in the record, as discussed above. One aspect to consider about palimpsests at the erosional level is that they tend to provide a false impression of the record. This can occur in arid areas where deflation exposes materials to a more visible level, either by augmenting the amount of artifacts on the surface, giving the impression that there are too many sites in the landscape (Fanning et al. 2009).

The importance of recognizing palimpsests in the record concerns particularly the problem of dating deposits, artifacts, and their contexts. Cumulative palimpsests can have residues that constitute legacies from older biochemical deposits such as organic matter, which can provide erroneous ages. For this reason, palimpsests should be considered a problem of legacy, and should be more discussed in the literature, and problems for their understanding and proper dating should be addressed, whether using simulations or by an extensive mapping of layers over an area.

MODERN ANALOGS, REFERENCE ANALOGS, AND MODERN REFERENCES

Practically, any science that uses proxy data to reconstruct conditions of the past relies on modern analogs. Despite their usefulness in representing processes in qualitative or quantitative form, modern analogs can entail several issues, from their conception to their application and interpretation.

In geology, and in particular in sedimentology, the idea of analogs is strongly embedded in the study of depositional environments (i.e., eolian, fluvial, deltaic, lacustrine) and facies models, which are often used to interpret environments using quantitative and qualitative sedimentological parameters (bedding, structure, chemistry, particle size distribution, etc.) trained in modern analogs (Canti and Huisman 2015). This idea is carried into geoarchaeology via Quaternary geology, to which the archaeological component is incorporated.

In paleoecology, the concept is applied to models developed on environmental parameters recorded from modern living organisms, as is the case of models developed for reconstructing vegetation and climate using pollen data, phytoliths, and diatoms. Although not done in geoarchaeology, the results of such models are often used as environmental background to geoarchaeological reconstructions.

Although the term modern analog is not frequently used in archaeology, the idea exists particularly in ethnographic and experimental research. Both fields study modern cultural processes such as the reproduction of artifacts and the formation of deposits, and in general analogs that explain the formation of the archaeological record. In environmental archaeology the use of modern analogs is largely common particularly in paleoecological studies. But overall the use of the environment is largely reproduced as a multi-relational approach as is the case of the now widely spread experimental farms. These ideas have permeated geoarchaeological thought and method, thus actualistic approaches such as experimental archaeology and ethnogeoarchaeology are using analogs of modern processes, but to answer questions related to aspects where geoscience research is involved (French 2003; MacPhail 2016; Tsartsidou 2016).

But despite the many intellectual merits and practical advantages in the interpretation of the record, modern analogs have received criticism, particularly in experimental research where modern conditions are assumed to be like those in the past (Goldberg and MacPhail

2006). This criticism also exists in paleoecology, where it has a strong basis in the fact that some Pleistocene ecosystems have no modern analogs, as evident in that many of the so-called natural ecosystems of today have been directly or indirectly transformed and influenced by human activities (MacDonald 2003).

The idea of the modern analog has its origins in the principle of uniformitarianism in geology, which has been debated and criticized on the basis of differences between past and modern conditions on earth. This criticism, however, reflects a poor understanding of the logical processes involved in the use of the present to interpret the past (Baker 2014). A close look at this process also suggests that, although it uses analogy, it does not necessarily use the idea of the present to reconstruct the past, but as a means of interpreting the record using a known *reference* of an observable process or depositional environment.

In reality, however, modern analogs serve as references, but they also serve as ways of calibrating models using measurements made in the present, which may not necessarily mean today, but recent historical times. Historical ecology, one of the fields that aims at providing a baseline for restorations, considers the idea of "reference conditions" (Egan and Howell 2001, 10). Therefore, the idea of modern also implies an idea of *reference* in the sense of a control sample, regardless of being obtained from a modern system. Thus, if the word "modern" is replaced by the word "reference," the idea of modern biases in interpreting past environments can be removed – a simple matter of semantics. Alternatively, if the word "analog" is replaced by "reference," the idea of reasoning by analogy can be removed.

In view of the above discussion, a modern analog can be interpreted as a *modern reference* in situations where it is imperative to stress the idea of observation in the present, as is the case of the interpretation of sedimentary facies models, which are based on modern parameters observable in streams and compared with the sedimentological properties of sediments and the architecture (e.g., Figure 3.5). In paleoecology a modern reference can be assumed to be a *reference ecosystem*, implying that a modern ecosystem is used not as modern analog but as a reference to reconstruct a model. Likewise, experimental farms (e.g., the Butser Farm in England) are modern references that help solve problems in the interpretation of human activities and natural processes of site formation, transformation, and destruction (Goldberg and MacPhail 2006).

On the other hand, a modern analog can be interpreted as a *reference analog* when the stress is put on the idea of a control sample to calibrate a model, without the implications of modern conditions, which can be seen in paleoecology with the idea of plant functional types (PTFs) or in cases where models are meant to explain trends or responses in probabilistic models, and using quantitative analyses, which could be seen in geomorphic responses to climatic variables.

SAMPLING AND INTERPRETATION OF THE RECORD

Although not directly seen as an issue of interpretation, sampling can influence the way record is interpreted. In other words, conclusions are based on results, and results are influenced on the resolution and extent of sampling. Therefore, sampling is an important aspect that needs to be managed during the early stages of the project, because in many situations re-sampling may not be possible.

Most archaeological and geoarchaeological manuals advise consideration of the sampling strategy from the beginning of the project, that is, during the design (Black and Jolly, 2003;

Carmichael, Lafferty III, and Molyneaux 2003; Goldberg and MacPhail 2006). It is then expected that sampling should be determined by the questions the research is trying to answer or the hypothesis of the work (Carmichael, Lafferty III, and Molyneaux 2003). However, in practical terms the actual sampling strategy may take turns depending on a series of circumstances, such as accessibility, budget, and unexpected turns in the research.

Technically, sampling takes place in the field during the stages of survey and excavation, corresponding in the geoarchaeological tasks as sampling on- and off-site (Figure 1.2). But special situations may dictate otherwise, particularly in cases when additional sampling is needed, or if sampling is to be done using indirect methods (e.g., remote sensing). Well-established projects with multiple, consecutive field seasons have the luxury of several rounds of sampling, sometimes allowing improvements along the way. The recommended rule of thumb is that on-site geoarchaeological sampling is done based on archaeological criteria, often stratigraphic, and archaeological sampling should follow aspects advised by geoarchaeological judgment. Similarly, it is often the unwritten rule that samples are split for various analyses, including for those out of geoarchaeological research (e.g. macrobotanical remains and pollen).

Most archaeological manuals provide prescriptions about sampling based on statistical principles, as for the use of random, stratified-random, systematic, and stratified sampling (Camichael, Lafferty and Molyneaux 2003; Renfrew and Bahn 2004). Resolution and sample size depend on the questions to answer and the nature of the deposit or floor. Systematic and random sampling also can be done at different scales during archaeological survey (Wells 2001).

In geoarchaeology, sampling can involve random collection, or representative samples of each layer, or systematic sampling of sections, that is to say, at determined intervals. But sometimes systematic sampling be problematic if layers are difficult to sample or do not produce relevant information, as is the case of gravel layers (Goldberg and MacPhail 2006). The situation also occurs horizontally, when floors are sampled for certain proxies. In certain situations, geoarchaeologists collect samples following soil or sedimentological techniques, but more commonly the sampling incorporates aspects related to the archaeological questions to solve (Courty 2001; Holliday 2003).

But beyond all the practicalities, obstacles, and circumstances, sampling is absolutely necessary for the interpretation of the record. Therefore, any interpretations of results should reconsider the way sampling was done and how representative samples are from the object in study. This retrospective analysis should be done in order to keep interpretations within the extent of the data and feed into future work if sampling strategies need to be improved.

CORRELATION AND ITS ISSUES

The practice of correlation

Throughout this chapter it has been stressed that causality is a fundamental aspect to explain phenomena that form the geoarchaeological record, and that the fundamental aspect of causality lies in determining the events that link cause and effect. This practice is common in sciences that deal with time-dependent phenomena such as geology and archaeology, or tend to look at events across time, a practice that stresses the relationist view of time discussed at the beginning of this chapter.

The relation between events in the stratigraphic record is carried out in the form of correlation, which is a common practice, but one that can have some issues that if not fixed can distort the real relation between events. Therefore, analyzing what correlation means and understanding its advantages and limitations is a safe practice in the reconstruction of events at all contextual levels.

Correlation is broadly defined in geology as the establishment of equivalence and corresponds between stratigraphic units. Although most dictionaries stress the stratigraphic component, others also suggest that phenomena can be correlated (Bates and Jackson 1984). In archaeology correlation takes a meaning that goes beyond stratigraphy, because the idea of cross-correlation between findings and texts has influenced the usage of the term. Furthermore, correlation can also be extended to linguistic, ethnographic, ethnohistoric and other anthropological, sets of data.

Judging by the usage in the literature, the idea of correlation also means finding equivalents between events, which, like geology, can be between stratigraphic sequences, or, like archaeology and history, between events. This implies correlation not only between cultural events, but also between cultural (or social) and natural (environmental) events.

Correlation of local events with climatic changes at regional and global scale is a common aspect that characterizes archaeological explanation of our times (Contreras 2016), and unfortunately one that can lead to erroneous conclusions, particularly because aspects of causality are ignored. Some of the causal aspects between social and environmental phenomena can be seen in historic and even in current times. For this reason, this book makes the point of using examples such as the devastation of New Orleans by Hurricane Katrina in 2005 (Case 13.2, Chapter 13) and the Dust Bowl of the 1930s (Case 14.2; Chapter 14) to show how complex correlations are across scales, even when all the information is at hand.

From local to global phenomena

In geoarchaeology correlations with the global phenomena approached when trying to find climatic explanations for certain events in the paleoclimatic record, as is the case of the Younger Dryas, the 8.2 ka event, the Medieval Climatic Anomaly (MCA), and the Little Ice Age (LIA), to give some examples. In other cases, certain disruptions in cultural evolution, such as collapse, are correlated with fluctuations of certain phenomena, such as the intensity of the Indian Ocean Monsoon, or with short-term oscillatory phenomena, such ENSO (El Niño Southern Oscillation), NAO (North Atlantic Oscillation), or the IOM (Indian Ocean Monsoon).

As a practice, particularly when various high-resolution independent proxies and data sets are used, correlations are possible. But explaining them in terms of human responses is difficult. To illustrate this point one has to look at the climatic changes in the contemporary past. At the moment, for example, the world is going through a trend to warmer temperatures, and one may assign certain events (e.g., hurricanes and droughts) to the result of them. But the matter is complicated in the sense that a single event may be not enough to support that assignment. To illustrate this situation, we can look at the devastation of New Orleans by Hurricane Katrina in 2005, and how the event might be seen by archaeologists 1000 years from now. Could this single event be related to the global increase in temperatures (i.e., the Hockey Stick), which would be recorded in global temperature records? Many meteorologists today would disagree with tying it to a trend that has developed over several decades (Knutson et al. 2010). If this trend had any influence on hurricanes, it is in the frequency of category-5 hurricanes, of which Katrina was one (Henson 2014). Certainly, Hurricane Katrina left an impact in the geoarchaeological record (see Case 13.2), but it will be very hard to link it to any global trend in temperature or even greenhouse gases. It will probably be necessary to scrutinize the interpretation with records in context, say by looking at the effect of hurricanes over a long period of time.

The effect of climatic oscillations, temperature trends, and long-term global phenomena can certainly leave an imprint in the geoarchaeological record, but other phenomena can also alter the imprint. At the moment, as the current climate change leaves its imprint in the geological record, so does human-led environmental change. So, distinguishing the causes of events in the record should then precede attempts at correlation.

5

The Human-Environmental Tradition in Geoarchaeology

INTRODUCTION

The human-environmental approach in geoarchaeology, the central theme of this book, originated from the need to integrate the broader environment into the geoscience-based approaches in archaeology. The theoretical model that allowed such an integration is the concept of *ecological context*, devised in various publications since the mid- to late 1970s and formulated in detail in Karl W. Butzer's (1982) treatise *Archaeology as Human Ecology* (AHE).

Since the publication of AHE, the ecological contextual approach has evolved along new paradigms developed through different geoarchaeological projects around the world and the influence of developments occurred in the science-society interface that has characterized the past three decades, particularly the increasing awareness of rapid global change.

This chapter is intended to briefly review the epistemological and historical basis of the ecological contextual approach in geoarchaeology and its evolution into the twenty-first century, stressing those aspects of research that are shaping the current thought and methods in geoarchaeology, particularly in the human-environmental approach. In doing so, this chapter becomes the theoretical framework for the subsequent chapters, which cover an array of traditional and emerging approaches in the study of human-environmental relations in geoarchaeology.

THE ECOLOGICAL PARADIGM

Ecology is the study of the interactions between living organisms and their environments (Sutton and Anderson 2010). These interactions are studied within the framework of the *ecosystem*, a concept introduced in 1935 by British ecologist Arthur Tansey to represent the interactions among biotic and abiotic components of a given environment (Lomolino 2010). This concept also portrays the idea of energy flows originating from solar radiation and circulating through the different biotic and abiotic components (Simmons 1996).

In its original form, ecology was closely related to the life sciences, where biologists, zoologists, and botanists use it as a framework to study living organisms in their environment. But since the mid-twentieth century the concept of ecology has begun to be adopted in the social sciences as a framework for placing humans within their broad environment, making ecology a paradigm in the broad spectrum of the study of humans (Finke 2014). Thus, fields such as human ecology, cultural ecology, political ecology, ecological anthropology, and ethnoecology began to appear as theoretical and methodological frameworks for studying human communities and their environments. In this process, in addition to *ecosystem*, ecological terms such as *ecotone, niche, habitat, energy, population, deme, carrying capacity,* and others were adopted in archaeology, anthropology, and other social sciences (Hassan 2004).

Human ecology first became part of sociology and then was used in anthropology, human geography, and other social sciences. But in reaction to several aspects of the place of humans particularly in traditional societies, the term *cultural ecology* was introduced by Julian Steward (1955), who proposed it as a method for studying traditional societies. Cultural ecology is also seen as part of the broader field of human ecology, which also includes biological ecology other social ecologies (Sutton and Anderson 2010).

In anthropology, criticism of Steward's view of cultural ecology lay on the fact that he saw humans as the center of the ecosystem and not merely part of it and presented a functionalist point of view (Moran 2007). Reactions modified the theoretical framework at the same time as other fields, such as ecological anthropology, part of environmental anthropology, and ethnoecology appeared (Crumley 1994; Townsend 2000; Moran 2007). These fields are, however, not replacements for one another, but different approaches within the broad ecological paradigm (*sensu* Finke 2014). Although many ideas and objectives overlap, each field has its own conceptual and methodological focus, some suiting the research in anthropology, others in sociology, and others in geography.

Soon after its original proposal, cultural ecology was adopted by geographers who reacted to the landscape morphology view of Carl O. Sauer (1925), which was criticized for being unscientific and having a "superorganic" (i.e., deeply anthropocentric) conception of culture (Robbins 2004). But in the latter part of the twentieth century, cultural ecology was criticized again by political ecologists who claimed, among other things, that culture is not always shaped by the environment, but by other social aspects as well (Robbins 2004; Moran 2007). Thus, it present most geographers have merged cultural ecology with political ecology to form *cultural and political ecology*, often referred by its acronym CAPE.

Archaeology has drawn concepts and models from ecology with the purpose of explaining human behavior in relation to the environment, or more properly placing human behavior in an ecological context (Dincauze 2000). In the 1970s, as geoarchaeology became a term in use and a field of study, the need for a contextual approach was evident, as geology did not provide the necessary link between culture and formation processes (Fedele 1976; Butzer 1975).

The ideas relevant to the ecological context in archaeology were further developed by Karl Butzer (1982) under the broad frame of human ecology (including cultural ecology), which principally incorporated the *multi-scalar* view from site formation processes to landscape and environmental processes. The ecological context also became an important theoretical concept in environmental archaeology, and in archaeology in general (Dincauze 2000; Branch et al. 2005).

THE ECOLOGICAL CONTEXT IN GEOARCHAEOLOGY

Ecological context as a multi-scalar approach

From its early days as a recognized field, geoarchaeology aimed at understanding processes of site formation, particularly in their archaeological and systemic contexts. The archaeological context has a broad meaning in terms of where the artifacts are found (e.g., primary or secondary contexts), and systemic context places the artifacts in terms of their behavioral value, which in turn is the basis for determining activity areas (Schiffer 1972). But as the ecological paradigm infused new ideas into the geoarchaeological approach, the idea of context began to expand to larger scales, particularly to an environmental scale (i.e., Butzer 1980a; 1982). *Ecological context* began to be used as a concept and paradigm, thus constituting the basis of the human-environmental approach in geoarchaeology.

In the ecological context, the landscape also plays an important role in the interpretation of the archaeological record. Although several conceptions of landscape exist in archaeology (see Chapter 3), in geoarchaeology the concept relies more on the geographic view, which not only includes natural and cultural features, but also a dynamic dimension, in many ways similar to that of the evolution of the landscape in geomorphology. The landscape is also an important contextual level to study aspects of land transformation and land degradation, as well as aspects of resilience, a trend that has put geoarchaeology at the center of the study of environmental change.

The cultural ecological framework and adaptive systems

The concept of adaptive systems as originally devised by Julian Steward (1955) has evolved. In the modern view, particularly cultural and political ecology, it is known as eco-cultural adaptation (Schubert 2015), which refers to complex societies, and in particular to contemporary societies. But in geoarchaeology the idea of adaptive systems has kept some of the original Stewardian ideas, particularly in terms of the diversity of cultural groups and environments and multiple interactions at different scales in time and space (Butzer 1982). In this sense, the idea of ecological context is extended to less complex groups such as early hominins, hunter-gatherers, and mixed-economic systems (semi-nomadic and pastoral).

The Butzerian contextual approach incorporates the ecological context of human cultures as *adaptive systems*, which have several measurable and replicable properties: space, scale, complexity, interaction, and equilibrium state (Butzer 1982). The properties of space, scale, and complexity are not much different from those discussed in previous chapters (e.g., Table 3.4). The idea of interaction is drawn directly from the concept of ecosystems, where biotic and abiotic components interact with one another. The idea of equilibrium state draws on concepts established in the 1970s by theoretical geomorphologists – concepts which, in Butzer's view, can be applied to human adaptive systems.

The adaptive system model varies according to different levels of social complexity, a term that denotes the degree of social and political hierarchy in a society. However, environmental influences can also add a degree of complexity to societies because the more complex the adaptive system, the more complex the relations with their environment.

Thus the concept of complex social-environmental systems (CSES) is proposed in Chapter 10, to dissociate complexity in the record with social complexity.

Paleoenvironmental information (e.g., climate, biota, soils, geomorphology, etc.) is necessary for reconstructing the functioning of the CSES. But as much as paleoenvironmental information is useful for putting adaptive systems in context, models of explanation are necessary to link environment with behavior (Schiffer 1987; Hodder 1999; Stein and Stein Dincauze 2000; Stein 2001). One such model is the Human Mode of Adaptation (HMA), which encapsulates the idea of environmental transformation based on the needs of groups to adapt (Dincauze 2000). The idea behind the HMA is that human groups face crises such as droughts or rapid environmental changes, which push them to develop new strategies.

Adaptive systems and the models to interpret them require different methodological models. As discussed in Chapter 2, methodological models have been provided by background disciplines. Each model entails a different methodological strategy for collecting information and interpreting depending on the society and environmental system in study. For example, hunter-gatherer contexts require a totally different set of methodological strategies than those, say, of agricultural societies. Likewise, the study of early hominin contexts in Africa requires different levels of abstraction than those, say, by late prehistoric foragers. Therefore, special adaptive systems have been developed by individual groups as means of tackling and discussing methodological problems, which are aspects discussed in the cases presented in Chapters 6 to 17.

Ecology and biogeography: A matter of scale

The framework of the contextual ecological approach in geoarchaeology cannot be seen as the direct influence of ecology alone. Although ecology in its different paradigmatic forms provides a context for biotic and non-biotic surroundings of humans, it fails to provide contexts at larger temporal and spatial scales (Jenkins and Rickfels 2011). For example, research subjects such as human evolution involve many phenomena and interactions with a long-term changing climate and with tectonics as well as aspects such as genetic variation, speciation, and variations among geographically separated populations (metapopulations), all of which are topics more related to the field of biogeography (MacDonald 2003; Lomolino 2014). Even certain processes of adaptation that represent shorter time periods, such as the case of plant and animal domestication, require a biogeographic explanation (MacDonald 2003).

Biogeography is the study of the spatial distribution of organisms and their ecosystems through time. Unlike ecology, biogeography takes into account the spatial aspect of living organisms. But, like ecology, it uses concepts such as ecosystem, population, community, carrying capacity, and predation, among others. The differences between ecology and biogeography are mainly in the spatial and temporal scales the two disciplines use for their subjects of study.

The concept of ecosystem, as used in ecology, does not necessarily portray the idea of spatial dimensions and boundaries. As an abstract model, the ecosystem does not have defined boundaries, and as an empiric model, its boundaries are usually arbitrary boundaries, either defined by a research plot, a reserve, or a study area. In biogeography, on the other hand, ecosystems have dimensions and boundaries (i.e., they are mappable) and have hierarchic categories according to scale. Thus, in biogeography the term *biome* portrays the same meaning as the term *ecosystem*, but it also portrays the idea of location, boundaries, and divisions based on variations with spatial distribution (e.g., ecozones, plant communities, or plant associations), transitions (ecozones), and changes over time (paleoebiomes). As an example, one can say "the tropical rainforest ecosystem," which does not portray the idea of a specific location, area, and limits. But, on the other hand, the "tropical rainforest biome" does portray the idea of its continental or global extension, as a mappable unit.

The separation between the idea of an ecosystem and a biome, among others, reflects the differences in the scales used by ecologists and biogeographers. Ecologists tend to study processes in short periods of time, often in confined parts of an ecosystem or in a research station. Biogeography, on the other hand, tends to look at processes over long periods of time and sometimes at larger scales. This difference leads also to slight differences in the use of common terms used by ecologists and biogeographers. For example, the "long-term" designation to an ecologist means more than 15 years, and perhaps up to 50 years; hence the term LTER (Long Term Ecological Research Station), which in the USA is a status given to those nature reserves and research stations that have carried out research for a long period of time, usually several decades.

To a biogeographer, "long-term" means a much longer time span, often involving thousands, if not tens of thousands, of years. Obviously, the latter has a more geological and historical connotation than the "long-term" idea of events in ecology. It is important, however, to bear in mind that although biogeography is more inclined to geological scales, it does have applications to explaining environmental archaeological processes.

Some of the most relevant biogeographic topics that geoarchaeology and in general archaeology benefit from include aspects such as biome shift, extinction, glacial *refugia*, domestication of plants and animals, dispersal, and even the more arcane Island Biogeography Theory. They all represent important topics in the ecological contextual approach and certainly in the interpretation of human-environmental relationships in the geoarchaeological record. In this sense, the ecological context, particularly at larger scales, often at the landscape and environmental level, can draw on biogeographic models.

Land use, land degradation, and related concepts

The landscape, at an ecological contextual level, receives significant attention in geoarchaeological studies, particularly in studies of land use change among other formers of land transformation and environmental change. Land use refers to the modification and management of a territory's ecosystems to adapt it to human needs. In other words, land use is seen as a replacement of the original vegetation and/or previous land uses with a new form of management. The functioning of any type of land use involves the flow of energy among biotic and non-biotic components, very much in the way it occurs in an ecosystem. It is for this reason that cultivated fields, pastures, artificial ponds, and other human-made land systems are referred to as *managed ecosystems* (Butzer 1994).

Linked to land use is the concept of land degradation, a term that often portrays the idea of human-led degradation, despite the fact that drastic climatic changes and tectonic and volcanic phenomena can degrade a landscape without the involvement of human agency. Examples of severe prehistoric soil erosion in the Mediterranean in pre-agricultural times show how important non-human forces can be on modifying the landscape (e.g., Van Andel 1998; Cordova 2000).

The term land degradation is often, and erroneously, used as synonymous with soil degradation, soil erosion, and desertification, among others. However, land degradation is a term that encompasses all forms of processes that impoverish the overall qualities of the land, including soil, vegetation, water, and even aesthetic qualities. Other terms, such as landscape change, landscape transformation, and environmental change as discussed below, may or may not entail land degradation. Therefore, by itself land degradation is only an aspect of other changes at the levels of landscape and environment.

Landscape transformation, environmental change, and environmental crises

At the contextual level of landscape, geoarchaeologists have been focusing on explaining processes of change such as erosion, sedimentation, the construction and/or destruction of features, alteration of hydrological and geomorphic processes, or any action that leaves a trace in the geoarchaeological record. However, the concept of landscape transformation does not necessarily encompass human processes that deteriorate the landscape (e.g., land degradation), but also processes of amelioration such as land reclamation and self-resilience (Cordova 2007).

Landscape transformation, one of the most popular topics in geoarchaeology, has a strong link to landscape archaeology, environmental archaeology, and it has become an essential methodological aspect in studies investigating society-environmental changes. In this sense, studies of societal collapse do have an important geoarchaeological component in a study of landscape transformation, as is portrayed in some cases in Chapters 10, 11, 14, and 15.

Environmental change is a topic of great intellectual interest in geoarchaeology and in the broader field of environmental archaeology (Branch et al. 2005; Rosen 2007a). Environmental change is defined as "departure from the mean or perceived state of any aspect of the environment" (Dincauze 2000, 63). In many cases, however, the trend is to link environmental change with climatic change, which is not always the case, as climate is only one of the factors at play in environmental change (Cumming, Cumming, and Redman 2006; Rosen 2007a; Butzer 2008).

The concept of *environmental crisis* refers to a particular event or series of events happening within the environmental history of a society, particularly in situations where the event leads to societal change. If the crisis disrupts an established social order, then the effect is what archaeologists refer to as societal collapse. However, not all environmental crises lead to societal collapse (Chapter 14). Also, an environmental crisis may be triggered by a natural disaster, but the disaster itself does not constitute the crisis. An environmental crisis can lead to an environmental disaster, but unlike catastrophes (e.g., natural disasters) it is a more protracted and complex phenomenon. To discuss these complex relations and the processes involved in them, natural disasters and environmental crisis are discussed separately in Chapters 13 and 14, respectively.

An environmental crisis can also involve complex relations between processes and events. The complexity lies in the multi-scalar and time-transgressive nature of such processes and events, which are sometimes difficult to discern in the geoarchaeological record. In many ways, the complexity of most environmental crises is related to social complexity, but in many cases to the complex relation between the society and the environment.

GEOARCHAEOLOGY SINCE *ARCHAEOLOGY AS HUMAN ECOLOGY*

After several decades, Karl W. Butzer's 1982 *Archaeology as Human Ecology* (AHE) is still cited and required reading in some archaeology and geoarchaeology courses. In addition to introducing many basic concepts and models to geoarchaeology, AHE strengthened the idea of the ecological contextual approach in archaeology and consolidated the basis of the human-environmental approach in geoarchaeology (Figure 1.1). However, in the decades since the publication of AHE, new concepts, ideas, and research methods have enriched the ecological contextual approach, as new environmental discourses have developed in society and science, and as recent technological advances have provided capabilities for more robust sets of data that facilitate the interpretation of the geoarchaeological record.

In the late 1980s and early 1990s, the emphasis on the human impact on the landscape became a popular topic in geoarchaeology, with many examples from around the world, particularly the Mediterranean region and the Americas (e.g., Brückner 1990; Van Andel and Zangger 1990; Denevan 1992; Turner II and Butzer 1992) and in a number of syntheses (e.g., Bottema, Entjes-Nieborg and van Zeist, 1990; Turner II et al. 1990; Lewin, Macklin, and Woodward, 1995; Dalfes, Kukla, and Weiss 1997; Leveau, Walsh, and Trement 1999). These studies in general began to address the issue of climatic vs. human causes in the transformation of the landscape, which in turn began to take a more global dimension, beginning to focus also on the broader issue of environmental change due to the ever-growing impact of humans on the environment.

In the first decade of the new century, the environmental discourses in archaeology and geoarchaeology began to explore more and more the issue of *societal collapse*. Although this topic had already become part of the archaeological and geographical discourses in the late 1980s and 1990s, it drew considerable attention in the early years of the present century, in part because of simplistic views of the problem in the popular literature (e.g., Diamond 1997; 2005), and in part because of the ever-growing awareness of the effects of human environmental impacts in the present as a mirror of similar effects in the past.

But not all the attention was given to the big picture of environmental change. Advanced technologies, particularly in the field of age determination methods, spatial information, remote sensing, and a number of other capabilities that permitted the collection, storage, and analysis of data also permitted new views into the formerly invisible parts of the geoarchaeological record as well as the possibility to explore processes across temporal and spatial scales (Gilhardi and Desruelles 2009). In essence, the new technological advances in spatial science (e.g., GIS and remote sensing) enabled the collection and analysis across scales, an aspect that defined in many ways the idea of ecological context in archaeology.

Towards the end of the first decade of the new century some of the studies of environmental change and past societies also began to look at new developments in the science of climate change and in the nascent concept of the Anthropocene, which have now become the two most important scientific frameworks to studying past human-environment interactions.

GLOBAL CLIMATE CHANGE AND THE RISE OF THE ANTHROPOCENE

The impact of global environmental changes on geoarchaeology

During the 1990s, several developments in the global society and environment contributed to the evolution of interpretive models created within the ecological contextual approach in the previous decades. One event that had a tremendous influence on the structure of models of environmental change in various disciplines was the Rio Declaration on Environment and Development that resulted from the Earth Summit convened in Rio de Janeiro in 1992. The influence of this event led archaeology and geoarchaeology to incorporate certain aspects of modern environmental concern to the study of past societies. Thus, emphasis on aspects such as sustainability, sustainable development, and vulnerable ecosystems, among others, were incorporated as frameworks for measuring the impact of past societies on their environments.

Another influential scientific and societal factor that shaped the study of past society-environment relations was the first of the IPCC (Intergovernmental Panel for Climate Change) Assessment Reports. The results allowed us to see, among other things, the human effects on climate and the global nature of weather phenomena, circulation patterns, and climate change. Additionally, the IPCC reports stressed the ocean-atmosphere-land interactions and reinforced the already existing idea of teleconnections among atmospheric phenomena (e.g., ENSO), providing a better way to look at causality in the case of climatic events of the past, which influenced the paleoclimatic approaches followed in the new century.

Also, in concomitance with the idea of ocean-atmosphere-land interactions were the increasing availability of isotopic data from deep-sea sediment cores and ice-sheet cores from Greenland and Antarctica, all of which began to provide high-resolution records of global climate change. The ice-sheet cores also provided significant information on the concentration of greenhouse gases in the past, opening the possibility to see that the impact of humans on climate predates the industrial era (Ruddiman 2014).

Needless to say, advances in climatology also provided tools to paleoclimatic models to better place other events, including societal events, in a climatic context. The new advances in paleoclimatology also raised awareness about the need for linking local and regional climatic phenomena with those at a global scale, and most important the new records revealed human influences on climate even before the industrial era.

Anthropocene arguments and the formation of a paradigm

The Anthropocene in general terms means the time in the Earth's history when humans had a significant impact on natural systems. Although the term has only recently been used and discussed in the literature, its history that goes back to the past century. The term was used in Soviet geological literature to mean the Quaternary, and it was not until the 1980s that German biologist Eugene F. Stoermer used it in the sense of the period when humans had a dominant impact on climate and the environment. Then the term was used sparsely in the literature in the 1990s, most notably in a book by Andrew Revkin (1994), and then popularized by Paul Crutzen in the early years of the century (Crutzen 2002). Since then, the Anthropocene has become the center of discussions in a number of fields, beginning with geology, and then with geography, the environmental sciences, and also archaeology.

The most important debate around the idea of the Anthropocene is its insertion as an epoch in the formal geological chronostratigraphy, an an idea hinted at originally by Andrew Revkin (1994). The problem is then when to place the beginning of the epoch. The general consensus expressed by the International Commission on Stratigraphy (ICS) marks it as the year 1945, as the times when radioactivity of the first nuclear explosions begins to be absorbed by the sediment record (Waters et al. 2016). It also marks the time when the amount of pollutants, particularly Pb, increased considerably in sedimentary deposits.

Nonetheless, not everyone agrees with the 1945 stratigraphic marker because it differs from the terms that sustain the stratigraphic markers, or Global Stratotype Section and Point (GSSP), that characterize the beginning of previous epochs, as is the case for the beginning of the Pleistocene epoch, which is marked by a sapropel layer overlain by shales in the deposits of the Mediterranean (Zalasiewicz et al. 2015). Similarly, millions of years earlier, the boundary between the last epoch of the Cretaceous and the first epoch of the Paleogene is identified by a layer of iridium, which marks what is known as the K-T event. In essence, the Anthropocene should be an epoch marked with an event of much larger proportions than radioactivity.

The other problem with having selected a particular point in time for the beginning of the Anthropocene is that human impact on the earth precedes the twentieth century. Furthermore, the human impact did not happen all at one time, but it has been incremental process since prehistoric times. So, independently from the stratigraphic marker established by the ICS, the debates go on as to when the significant anthropogenic effect began to be felt on the Earth's lithological, atmospheric, and biological systems.

Although Paul Crutzen in his original Anthropocene paper suggested several possible times for the beginning of the Anthropocene as a geological epoch, he and his colleagues in subsequent publications have proposed the beginning of the Anthropocene as around the beginning of the industrial revolution, that is, around 1850 (Crutzen 2002; Crutzen and Sheffen 2003). However, the industrial revolution is not an event that happened overnight. It took several decades to really be noticeable. In other words, seen in the atmospheric carbon record before the *Hockey Stick* (late 1970s, early 1980s) the industrial rise was steady.

Another argument against marking the beginning of the Anthropocene with the industrial revolution is that carbon dioxide (CO_2) and methane (CH_4), the two most abundant of the greenhouse gases, had risen from the normal course several millennia earlier since the beginning of agriculture (Ruddiman 2014). Forest elimination for planting crops, soil breaking and removal, and the creation of rice paddies have steadily contributed to the production of CO_2 and CH_4 (Ruddiman 2014). The problem in recognizing this impact is that when put against the rapid increase in greenhouse gases of the industrial revolution the contribution of agriculture in pre-industrial times is minimized.

Since its beginnings agriculture also had an impact on geomorphic processes that changed sedimentation records in areas where agriculture was practiced. Furthermore, the removal of vegetation is also marked in pollen records in lakes. However, such stratigraphic markers are more regional and spotty in time and space, and in some cases not easily perceptible in records. Therefore, the beginning of the Anthropocene is a time-transgressive phenomenon – the very reason it is difficult to pinpoint it in a global stratigraphic context.

Beyond its contentious stratigraphic definition, the Anthropocene has become the conceptual framework for measuring the degree of interaction between humans and their environment, a topic that has become central to archaeology and certainly to geoarchaeology (Butzer 2011, 2015; Edgeworth 2013). This can be seen not only in the *anthropization* of the landscape, but also in the modification of environmental patterns, both of which are an important aspect in the study of human-environmental interactions.

PERSPECTIVES ON THE HUMAN AND NON-HUMAN WORLDS

The idea of early human-environmental relations, now conceptualized in the idea of an Anthropocene, has a long intellectual history, even going back to ancient Greek philosophy. The concept of *Ecumene* (From Greek, *oikouménē*, lit. inhabited) portrayed the idea of a world inhabited by humans, a world presumably different from that uninhabited (Berque 2015).

In the twentieth century, the concept of inhabited world has been merged with the physical world in the context of the *Noosphere* (from the Greek, *nous*, mind, and *sphaira*, sphere), a term introduced in the literarture by Pierre Teilhard de Chardin (1881–1955), a French geologist, geographer, and anthropologist, who took on the paleoanthropological study of homins in China.

In his 1923 essay, *Hominization: Introduction to a scientific study of the human phenomenon*, Teilhard de Chardin (1957) defined the concept more in terms of human cognition originating from a complex biological evolutionary process, which were also discussed by the French philosopher Édouard Leroy (Bailes 1990). But it was up to the Russian-Ukrainian geochemist Vladimir V. Vernadsky (1863–1945) to redefine the Teilhardian idea of the noosphere in its spatial and temporal dimension, introducing the idea of noosphere as a new sphere that developed as humans began to interact and change the structure of the biosphere (Bailes 1990). The idea implied by V. V. Vernadsky is that the noosphere interacts with the lithosphere, atmosphere, hydrosphere, and biosphere, thus being one of the planet's spheres.

Interestingly, like the noosphere, the ecumene "emerged from out of the biosphere, mainly through the processes of hominization, anthropization and humanization" (Berque 2009: 159). In essence, this means that the human world was created by the process of becoming *Homo* (hominization), by occupying new lands across the world (anthropization), and becoming modern humans (humanization), with the latter being the most important in creating the ecumene.

The concepts of noosphere and ecumene have given way to other conceptions of the human world, for example, the Gaia hypothesis, publicized by James Lovelock and very popular among biologists and environmentalists. Unlike the ecumene, which is both a historical and geographic term, the noosphere and Gaia hypothesis are developed at a more conscious and abstract level with little ground on the time-space dimensional framework. However, the three of them are old concepts that in many ways relate to the conceptual framework of the Anthropocene.

In contrast to the ecumene, the noosphere, and Gaia hyothesis, the Anthropocene is a concept that involves the temporal and spatial dimensions of human presence in the world (i.e., the process of anthropization), which makes it more attractive to archaeologists, geographers, and geologists as a framework for the study of the increasing influence of humankind on the global environment. One problem for conceptualizing a universal view of the Anthropocene is that the place of humans in the world is sometimes seen differently, depending on the discipline and sometimes the different theoretical frameworks. This is an important aspect for geoarchaeology, where several influences (archaeological, anthropological, geological, geographical, and ecological) modify the way we study humans in the environment.

Humans as an event in the history of the world

The geosciences, and in particular geology, study the evolution of the planet. In the long history of the world, hominins occupy only a fraction. When humans enter the scene, social scientists and humanists join the common interest in the matter. Ironically, however, concepts such as the noosphere and Anthropocene were first proposed by geologists (e.g., Pierre Teilhard de Chardin, Vladimir I. Vernadsky, and Paul Crutzen).

The idea of humans as part of the history of the world are more associated with the process of anthropization, which contrasts with hominization (a more anthropological term) and humanization (a more humanistic, if not religious term). Pierre Teilhard de Chardin (1957) tried to elevated the humanistic role of humans in the noosphere, in part because he tried to reconcile science and religion. On the other hand, in his conception of the noosphere, V. I. Vernadsky placed humans at the same level of animate and inanimate things of nature. In this conception, human evolution was not much different from the evolution of other species.

The idea of humans as an event in the history of the world has prevailed in geology in part because of the time scales that the field often works with. But this situation is not extended to all areas of geology, particularly now that the concept of the Anthropocene is seen as an epoch. Some of these ideas may also have existed in other natural sciences, where the concept of the anthropic has been stressed as unusual. Nonetheless, it is important to acknowledge these conceptions because of the importance that geological thought has had in geoarchaeology.

Humans as an anomaly in the ecosystem

The idea that humans are an anomalous phenomenon in nature derives in part from the idea that humans evolved relatively fast and began to transform other species and the environment too quickly. These ideas had more influence in areas such as geography, biology, and environmental sciences, and less in anthropology and archaeology, which were more influenced by the social aspects of the ecological paradigm.

However, some ideas of the anomalous human have permeated some archaeological paradigms. Most notably these ideas in North America influenced the followers of the *Overkill* Theory, or the idea that humans exterminated the Pleistocene megafauna, which, among other things, was welcomed by many of the supporters of the then-in-vogue Clovis-First Theory.

The idea of humans as an anomaly of the ecosystem is ambivalent when it comes to assessing periods of time and place. In areas such as Africa, where humans evolved, it seems very clear that a co-evolutionary process existed between humans and the ecosystem. Not so, however, in continents and islands where humans arrived in a rather quick manner, particularly as anatomically modern humans (Simmons 1996).

However, the issue that is often debated is how to define the time when humans detached themselves from their equal status with other hominins. This in part has to do with the production of tools and management of fire, which was fully attained by *Homo erectus* (Simmons 1996). With this comes the question posed by the concept of Anthropocene: when did humans begin to impact their environment? As discussed in Chapters 6 and 7, evidence of pre-*Homo sapiens* impacts exists at certain scales. Nonetheless, the clear imprint does not occur until anatomically modern humans, or modern Homo sapiens appear in the scene.

Humans as part of the ecosystem

Because humans are now in control of practically all the world's ecosystems, many scientists believe that it is difficult to see the functioning of ecosystems in a purely natural environment. In other words, the idea is that modern or recent ecosystems function basically with the presence of humans, particularly in cases where humans have already transformed the flows of energy. The hypothetical scenario in Allan Weisman's (2007 *The World Without Us* seems to portray the idea that if humans suddenly disappeared from the planet, the world would enter a long phase of ecological chaos.

This idea of humans controlling the flows of energy in most ecosystems has been addressed in the archaeological and geographical literature (e.g., Dincauze 2000; Butzer 1996), where the concept of anthropic ecosystems or managed ecosystems is used, but not in the sense of total dominance. The original ideas proposed by the early influences of the ecological paradigm, notably with the case of the Stewardian cultural ecology, discussed earlier in this chapter, created the idea that humans are part of the ecosystem, and that humans evolved with it. The idea of the noble savage, for example, portrays this idea.

GEOARCHAEOLOGY AND ENVIRONMENTAL HISTORY

Using the methods and approaches of the geosciences, geoarchaeology contributes to the understanding of human-environmental relations in the past, a topic that is of interest not only in archaeology, but in other scholarly fields such as environmental history, an academic enterprise that has developed close relations with geoarchaeology (Butzer and Endfield 2011).

Environmental history is a relatively recent field, dating to back to the 1960s and early 1979s (Hughes 2015). For many social scholars, particularly in the USA, environmental history is a field of history, although it is purportedly practised by sociologists, political scientists, geographers, and archaeologists (Crumley 1994; Hughes 2015). The overlap between topics in environmental history and those of various disciplines makes obvious that it is not one field, but a multidisciplinary field.

Geoarchaeology is increasingly seen as an important tool to explain those phenomena that originally could not be explained only by history (Butzer 2011). There are numerous examples in environmental history that are beginning to draw on data produced by the sciences that work in tandem with archaeology, particularly geoarchaeology and paleoecology. Cooperation between geoarchaeologists and environmental historians can fit within the needs represented by the three dimensions of environmental history proposed by Hughes' (2008) three continua: the cultural-nature, history-science, and time-space continua. The potential contributions of geoarchaeology to the three dimensions is undeniable because in a way these dimensions coincide with some of the aspects geoarchaeology deals with, particularly with the cultural and natural processes and the different scales in time and space.

6

Geoarchaeology and Human Evolution

INTRODUCTION

Human evolution comprises a lengthy process, of at least six million years, during which extinct hominins evolved into modern humans. Hominins, short for Hominini, constitute a tribe that includes several genera that lived during the late Miocene and Pliocene, such as *Orrorin*, *Ardipithecus*, and *Australopithecus*, and during the Pleistocene, such as *Paranthropus* and *Homo* (Picq 2003). Of all the hominins, the genus *Homo* had by far the largest geographic expansion and is the only one that survived to this day as *Homo sapiens sapiens*.

During the lengthy process of human evolution, the earth underwent several geological, biogeographic, and climatic changes that directly or indirectly influenced the process of hominization, or the evolutionary change that led to the genus *Homo*. Some of the evolutionary traits in this process of hominization include *bipedality* (or *bipedalism*) and *encephalization*, (Coppens 1999; Gamble 2013). For this reason, the study of geological, paleoclimatic, and paleoecological changes are of importance, along with the research efforts to study hominin fossils.

The study of human evolution is led by paleoanthropology, a branch of anthropology that studies hominin fossils and their evolutionary relations to modern humans. Paleoanthropology in many ways is a link between biological anthropology and paleontology, a matter that is important to mention in this introduction to the chapter because of the ways paleoanthropological archaeology differs from other archaeological fields, particularly in terms of perceptions of the human subjects, scales, and methodological approaches (Figure 6.1). Accordingly, such differences influence the way geoarchaeology participates in the broad field of human evolution.

It is noticeable in the human evolution discourse that the term "human remains" is used to refer to later hominins, particularly those of the late *Homo* species, while the term "fossil remains" is applied to earlier hominin taxonomic groups (Figure 6.1). Thus, it is not uncommon that the earliest, usually pre-*Homo*, hominins are viewed at the same ecological level of mammal fauna, not only in biostratigraphic ways, but also in terms of biogeographic processes such as extinctions, migrations, and genetic bottlenecks (Leakey and Werdelin 2010; Harcourt 2012). Furthermore, the long time scales involved in human evolution affect the resolution at which climatic events are analyzed, focusing on oscillations that are more consistent with long-term orbital cycles and tectonic processes at continental scales.

Broad hominin groups →	Pre-*Homo* hominins	Early African *Homo* species	Out of Africa *Homo* species	Archaic *Homo sapiens*	Modern Humans
Scholarly field	Paleoanthropology Archaeology ———			Paleoanthropology Archaeology	Anthropology Archaeology
Hominin remains	Fossils ——— - - - - - -			- - - - - - ———	Human remains
Environmental approach	Biogeographical ———		Biogeographical & ecological		Ecological & biogeographical
Geographic focus	Africa	Africa	Africa and Eurasia	Africa and Eurasia	The World
Stratigraphic-time approach	More geological/deep time/ paleontological	Ocean-isotopic	Ocean-Isotopic	Ocean/ice-Isotopic	Ice-Isotopic
Litho-stratigraphic units	Formal lithostratigraphic (group-formation-member)		Formal lithostratigraphic and allostratigraphic	Mostly allostratigraphic	Allostratigraphic Ethnostratigraphic
Past-modern settings	Mostly unconformable	Variable		Variable	Mostly conformable
Time scales	Global-continental	Continental	Continental-regional	Regional -local	Regional -local
Environmental forcings	Tectonic Orbital	Orbital	Orbital	Ocean-atmosphere Orbital	Ocean-atmosphere Greenhouse
Age/Time scale in years	>3.3 Ma/10^6	3.3 - 1.8 Ma/10^6 - 10^5	1.8-0.25 Ma/10^6 - 10^5	250-50 ka/10^5	50-0 ka/10^5 - 10^1

Figure 6.1. Spectrum of conceptions and perspectives on the archaeological record of hominins depending on age and taxon.

The long time frames involved in human evolution also determine the way geoarchaeology is applied to the reconstruction of early hominin environments. Thus, the stratigraphic frameworks used in the study of early hominins in Africa are more characteristic of deep time geology (Butzer 1978; Rapp, Jr. 1981; Brown and Feibel 1986; Grayson 1990), which often use the formal lithostratigraphic units at the level of group, formation, member, and bed (Figure 6.2) as opposed to the allostratigraphic units used in later prehistoric times.

In conjunction with the stratigraphic schemes, most hominin sites, particularly in Africa, are associated with geomorphic settings that are unconformable with modern ones. Thus, the lacustrine or alluvial settings of most hominin sites in Africa have been deeply modified tectonics, volcanism and erosion, to the point that in some cases former lacustrine and alluvial surfaces have been inverted, now occupying upland settings, or in some cases steep canyon slopes. Paleogeomorphic settings are thus found only through the study of sedimentary facies much in the way it is done in deep-time geology.

In terms of space and location, the study of early hominins also represents a special geographic component in human evolution geoarchaeology. First, Mio-Pliocene and Plio-Pleistocene hominin sites are restricted to Africa, and in particular to eastern and southern part of the continent. Most early fossils belonging to the genus *Homo*, particularly Homo erectus, were restricted to Africa and the southern half of Eurasia. This disproportion of time vs. global space is illustrated in the map sequence of Terrae proposed by Clive Gamble (2013) (Figure 6.3).

Practically, it is only at the end of Terra 3 (ca. 50 ka) that hominins, now in the form of modern Homo sapiens, begin to colonize the north of Eurasia and other continents

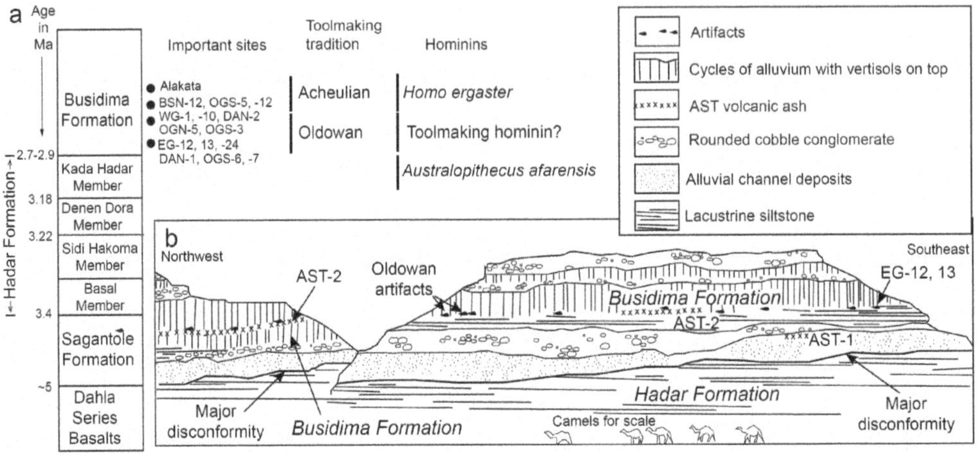

Figure 6.2. a) Stratigraphic scheme in the Afar Area, Ethiopia; and b) sedimentary facies of the Budisima Formation with associated Oldowan artifacts. Modified from Quade et al. (2004).

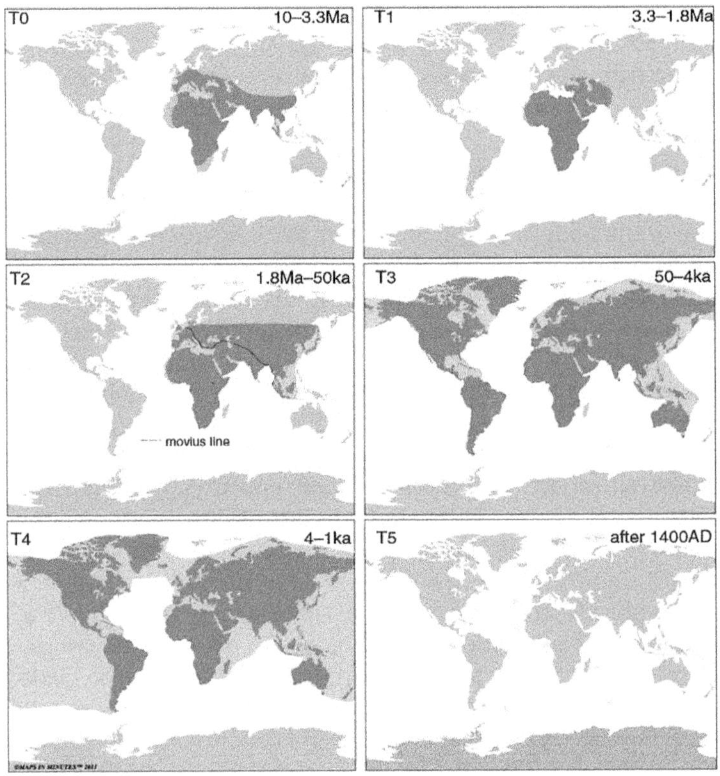

Figure 6.3. The Terrae according to Gamble (2013), showing the extent of pre-*Homo* hominins (T1-2), early *Homo* species (T3), and *Homo sapiens* (T4-5).

(Figure 6.3). This restricts most of the research on human evolution to Africa and southern Eurasia. This geographic distribution and all the geological and biogeographic aspects linked to the evolution of hominins make geoarchaeological work in the field of paleoanthropological archaeology unique. It is then important to discuss briefly the geologic, climatic, and biogeographic contexts of human evolution, and discuss the role of geoarchaeology in paleoanthropological research.

GEOLOGY, CLIMATE CHANGES, AND BIOGEOGRAPHY

Early hominin contexts: Archaeological or geological stratigraphy?

During its early days, archaeology borrowed from geology the principles of stratigraphy. Thus, it is not surprising that for a long geology and archaeology had common roots (Hodder 1999; Rapp, Jr. 2007). Eventually, however, archaeological stratigraphy veered away from geological stratigraphy, as archaeologists needed different models to interpret the complexities of cultural deposits (Harris 1989). But oddly enough, because of the time frames involved in the hominin record, the interpretation of stratigraphy in paleoanthropological archaeology, particularly those in the Miocene and Pliocene, draws on stratigraphic approaches more akin to geology than archaeology. As shown in the diagram of Figure 6.1, hominin fossils are seen up until some point, roughly in the Middle Pleistocene, as paleontological remains along with other fauna, sometimes even constituting an important aspect in relative dating.

Given the stratigraphic aspects of hominin fossils, geological fields such as geochronomoly volcanology, sedimentology, and paleontology become necessary in aspects ranging from dating to paleoecology (Garrison 2016). Structural geology should be added to the interpretation of stratigraphic and geomorphic contexts of hominin remains because during the long time spans of human evolution, faults and other tectonic deformations influenced the sedimentary record, a situation typical of the East African Rift.

However, the strength of geological scales prevailing in the interpretation in early hominin records does not undermine the role and essence of archaeological stratigraphy. At small scales, hominin behavior can be studied in cases when artifacts, modified faunal remains, hearths, and chemical traces are in a primary context, as in the example discussed in Case 6.1.

The long-term trends of warming and cooling

Climate and vegetation changes in the early evolution of the subfamily Hominidae and the tribe Hominini allegedly played an important role, particularly in bipedality and encephalization (DeMenocal 2004; Potts 2007; Gamble 2013). These environmental changes may, in turn, have contributed to dietary changes, and eventually to the evolution of culture, which among other things may have been important in the eventual dispersal of late hominins (genus *Homo*) across and out of Africa (Leakey and Werdelin 2010).

Climatic changes were in large part forced by variations in insolation caused by orbital changes (i.e., Milankovitch Cycles) and to a lesser extent and more regionally and locally by tectonic changes (Maslin et al. 2014). But both long-term climate changes and tectonics

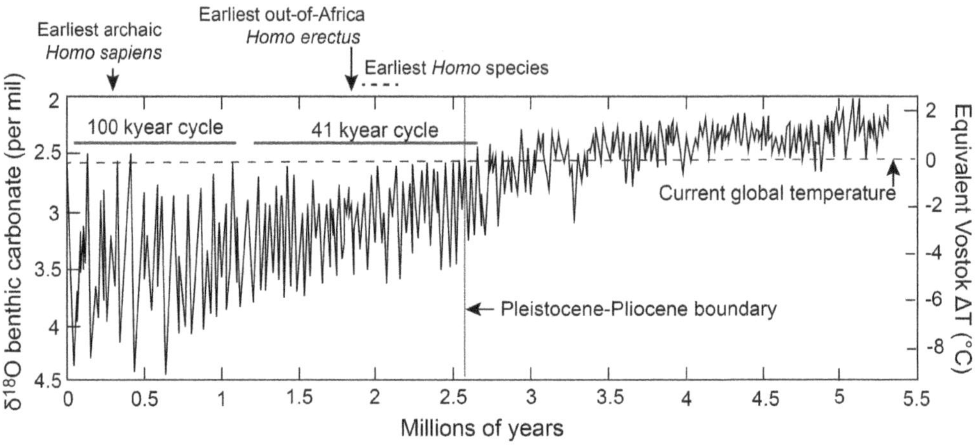

Figure 6.4. Temperatures for the past five million years obtained from isotopes. Composite figure using data from different sources, particularly Maslin et al. (2014).

influenced the evolution of terrain, hydrology, flora, and fauna, all of which were part of the environment where hominins evolved.

The global climatic change background in human evolution is usually interpreted from δ^{18} records from deep-sea cores (e.g., Figure 6.4) and regionally from paleosol and fossilized dental $\delta^{13}C$ records (DeMenocal 2004; Lee-Thorpe, Sponheimer, and Luyt, 2007; Quade and Levin 2013); and lacustrine sediments, particularly in the East African Rift zone, can also provide valuable information on regional dry and wet trends in response to global temperatures (Maslin et al. 2014). Additional local-to-regional records exist in shorter sequences from pollen, phytoliths, diatoms, and other proxies, which, though discontinuous, provide a sequence of short-term windows that show the influence of climate on hominin populations and their environment.

The general trend of global temperatures from the Mio-Pliocene through the Pleistocene is towards cooling, with downs and ups corresponding respectively to glacials and interglacials (Figure 6.4). For Africa, the cooling that the northern hemisphere experienced during glacial periods meant arid conditions (DeMenocal 2004). But not all glacial and interglacial periods were the same, since they changed in amplitude and duration. In fact, there were longer periods in which shifts occurred in the mode and duration of glacial-interglacial cycles.

During the Pliocene and Pleistocene at least three important events changed the amplitude of the glacial-interglacial cycles, marked by three transitions (Figure 6.4). These transitions include the intensification of the Northern Hemisphere glaciation (2.7–2.5 Ma); the intensification of the Walker Circulation of the tropical oceans (1.9–1.7 Ma); and the early Mid-Pleistocene revolution (0.9–0.7 Ma) (Maslin et al. 2014). The first transition caused a cooling of the oceans and an increase in African aridity; the second transition encompassed an increase in cooler tropical oceans with more evidence of phenomena such as El Niño Southern Oscillation (ENSO); and the third transition marked a stronger amplitude in the glacial-interglacial cycles (Figure 6.4). These transitions have been recorded in the lakes and in the numerous proxies, particularly $\delta^{13}C$ from faunal remains and soils (Maslin et al. 2014).

The role of biogeography in human evolution and hominin migrations

Climate change is not only a background to the interpretation of the geoarchaeological record of hominin sites, settings, landscapes, and environments, but also a source of information on the geography of the hominin species, a field that is properly called human biogeography (Harcourt 2012). Global climatic changes directly affect the geography of hominins, the process of hominization, and the migrations (MacDonald 2003; Finlayson 2005; Harcourt 2012). Drops in global temperatures not only altered the climates of Africa and other low-latitude areas, but also locked moisture in ice, changing coastal geographies, and allowing passage from lands that today are separated by sea (MacDonald 2003). Thus, concepts related to glacial-interglacial transitions include biogeographic concepts such as dispersals, land bridges, filters, glacial refugia, genetic bottlenecks, and in general co-evolutionary relation among flora, fauna, and hominins, all of which have been crucial to model processes of human evolution (Harcourt, 2012; Dennel et al. 2014; Carotenuto et al. 2016). Although these topics seem disconnected from geoarchaeology, they do represent the theoretical basis of environmental models that help shape geoarchaeological research strategies and interpretations of the record.

In connection with climatic changes is the concept of *glacial refugia*, now widely used in archaeology and paleoanthropology, is borrowed from biogeography (Cordova et al. 2013). Despite being "glacial" in its biogeographic form, the concept does not necessarily refer to glaciated areas, but also to the form of oases in dry areas or patches of tropical rainforest that survived in Africa during glacial aridization periods (MacDonald 2003). The case of desert refugia is an important factor in the migrations of humans across the world during adverse climatic conditions, as is the case of the dispersals of hominins out of Africa, which happens through bridges across extremely arid areas of the Sinai and the Arabian Peninsula (Bar-Yosef 2000; Armitage et al. 2011). Thus, the concept is also tied to hominin migrations, which in biogeography corresponds to the concept of dispersals.

Models of dispersal have been topics of great interest in modeling the explanation of the waves migrations of *Homo* species, in particular the movement of *Homo erectus* out of Africa (Harcourt 2012; Carotenuto et al. 2016), an event occurred around or earlier than 1.8 Ma. Of importance is also the migration of modern *Homo sapiens* out of Africa, across Eurasia and into other contients in the Late Pleistocene (Finlayson 2005; Carotenuto et al. 2016).

Although little is known of movements of previous hominins across the Africa-Southwest Asia bridge, information on the most recent passage by modern *Homo sapiens* has been more closely studied particularly in the southern part of the Arabian Peninsula. This region, although extremely arid and not connected by land to Africa today, represented phases of opportunity in terms of low sea levels and relatively short wet periods linked to the intensification of the Indian Ocean Monsoon (Armitage et al. 2011). The intensification of monsoon rains had raised precipitation in the southern part of the Peninsula (Petraglia and Alsharekh 2003; Breeze et al. 2016). The rest of the sites in the central and northern part were influenced by local conditions, particularly oases – the basis of desert refugia. Many of the Middle Paleolithic sites in this region are associated with refugia near springs and paleolakes (Figure 6.5). Farther north the influence of wet periods of intensified Mediterranean cyclogenesis may have also constituted times to link such refugia spots.

Although the biogeography topics seem disconnected from geoarchaeology, they do represent the theoretical basis of environmental models that help shape geoarchaeological

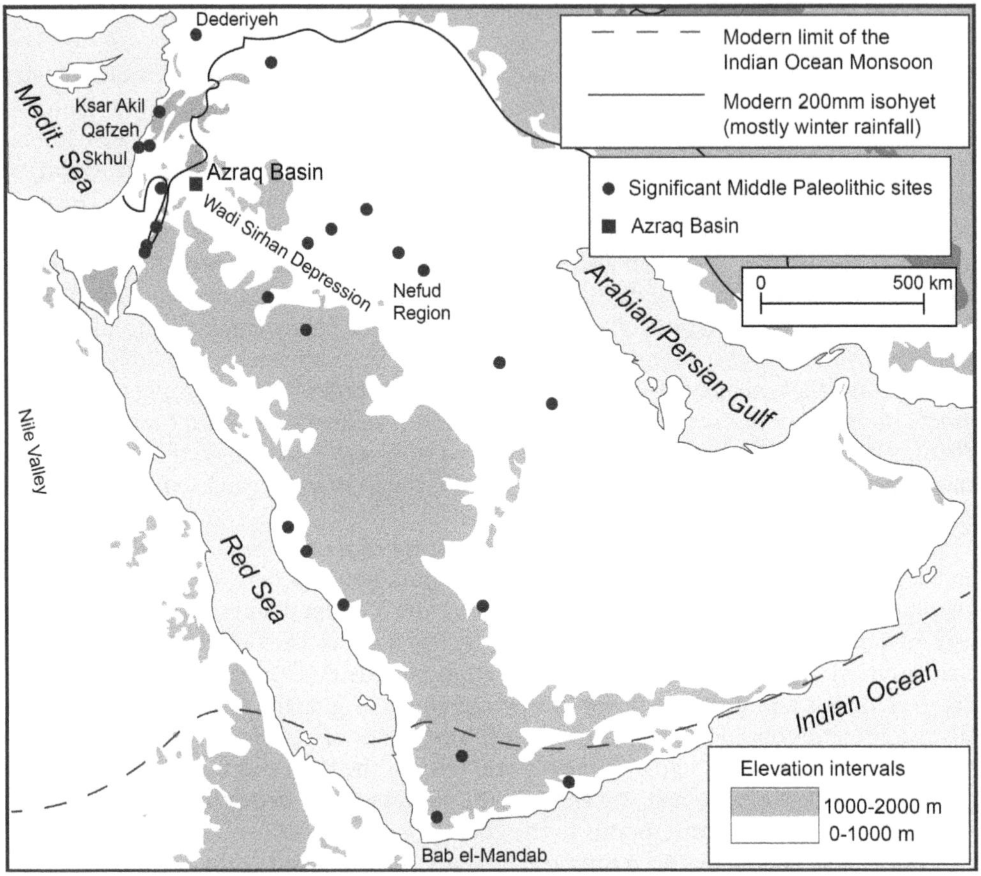

Figure 6.5. Arabian Peninsula with significant Middle Paleolithic localities dating probably to the time of passage of anatomically modern humans from Africa (modified from Cordova et al. (2013), with information from Petraglia and Alsharekh 2003).

research strategies and interpretations of the record. Furthermore, biogeographic models are key for the interpretation of long-term human-environmental relationships.

GEOARCHAEOLOGY IN PALEOANTHROPOLOGICAL RESEARCH

The role of geoarchaeology in the broad study of human evolution is essential in aspects such as site prospection, geochronology, and paleoenvironmental reconstruction (Butzer 1978; Rapp, Jr. 1981; Grayson 1990; Stern 2008; Feibel 2013). Studies of geomorphology and sedimentary systems provide information on landscape change, which may be related to major geologic and climatic transformations (Butzer 1978; Rapp, Jr. 1981; Grayson 1990; Feibel 2013).

However, it is important to recognize that geoarchaeological research faces several challenges in the study of early hominin sites. One of them is the time scales involved, which, as explained above, encompass large periods of time. The problem is that, often, paleoanthropological research needs higher resolution in certain aspects, which is difficult to find. Therefore, drastic changes from scales encompassing site formation processes to those explaining long-term climatic trends and tectonics requires a look at different parts of the record over large areas (Butzer 1978).

Additional to the broad time-scale amplitude one has to consider issues with the fragmentation of the record, which may be caused by erosion and lack of sedimentation. Incomplete records can result from gaps created by large areas that have not been studied due to poor accessibility or because they straddle several countries and areas of political conflict.

Despite the limitations associated with time scales and a fragmentary record, geoarchaeology aims at finding strategies for recovering information valuable to paleoanthropological research. This task requires the understanding of processes that form the record of hominin sites, settings, landscapes, and environments. However, as is usually the case in paleoanthropological archaeology, there are a number of other specialists working at the same contextual levels, which implies then coordination and share of information.

CONTEXTS AND ISSUES

Earliest hominin sites and settings: Environments and long-term landscape change

The earliest hominin sites, roughly corresponding to Terra 2 (Figure 6.3), are for the most restricted to the eastern and southern parts of Africa (Figure 6.6). In these regions, geoarchaeological work has been focused on numerous tasks, particularly the reconstruction of geological and climatic changes that occurred in the late Miocene, Pliocene, and early Pleistocene (Quade et al. 2004; Reynolds, Bailey, and King 2011; Maslin et al. 2014). These reconstructions, however, include data obtained at all contextual levels.

Studies at the level of site are difficult to define in some context where the record is widespread, a problem that also occurs in later hominin contexts, and even in the study of hunter-gatherers (see discussion in Chapter 8). However, great prospects for microstratigraphy and reconstruction of behavior are possible, as shown in Case 8.1. But there are many other studies that are often included as site-setting, which is significant in terms of dating, formation processes, site taphonomy, and prospection of new sites.

Geomorphic settings are varied depending on the area of study. In the East African Rift most sequences involve lacustrine and alluvial sediments with beds of volcanic ash. The latter have been useful in dating the finds using K-Ar, a method that gives more resolution than paleomagnetism. In other locales, particularly in southern Africa, deposits in karstic cavities in dolomites are highly common, as is the case in Taung, Molopa, Sterkfontein, and Swartkrans, where most deposits are consolidated into breccias (Figure 6.6).

Taphonomic study of the hominin record is an important issue that requires geoarchaeological techniques, particularly at the site-setting level, and even in a broader sense at the level of landscape. For example, very common it is observed that upland areas are void of artifacts, while the rapidly aggrading deposits contain more, suggesting that perhaps upland areas were never settled (Butzer 1978). However, intense erosion in the uplands is an issue that better explains the scarcity of materials in uplands (Butzer

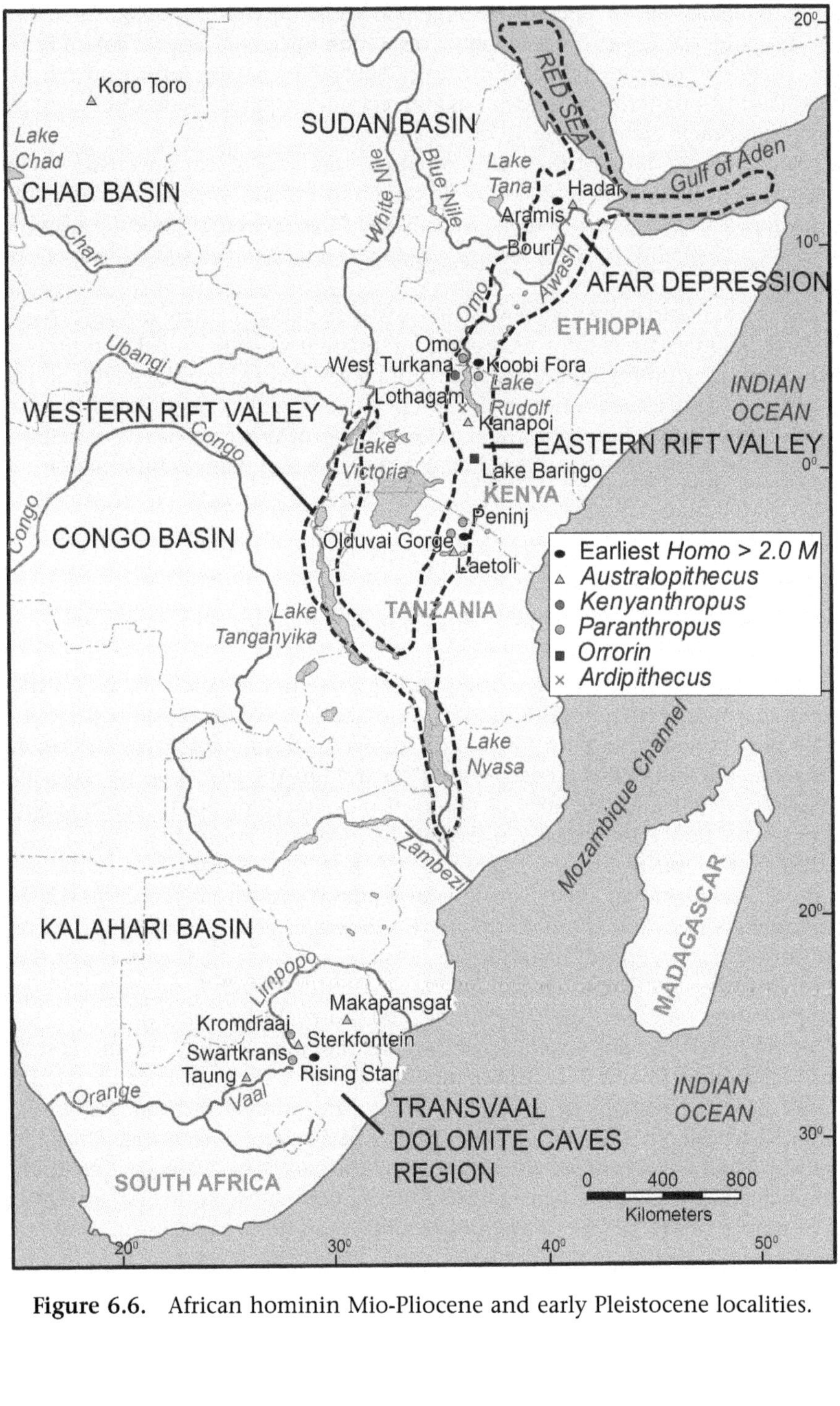

Figure 6.6. African hominin Mio-Pliocene and early Pleistocene localities.

1978; Reynolds, Bailey, and King 2011). Therefore, taphonomic issues at the level of setting and landscape have to be considered in the interpretation of hominin paleoenvironments.

Site prospection in general involves studies of settings, which may seem either recurrent in certain landforms or stratigraphic layers. As mentioned above, the unconformity of settings is one problem to solve, which in turn represents a resource for the reconstruction of paleolandscapes. Site prospection in the study of early hominin sites is another issue related to geographic concentration of sites in particular physiographic provinces, whose explanation requires a broad understanding of geology, climate, and biogeography at different temporal scales.

Geographic distribution of early hominin sites

With the exception of *Australopithecus bahrelghazali* in Koro Koro, Chad, on the southern fringes of the Sahara Desert, all pre-Pleistocene hominin remains have been found in these two regions: the East African Rift (EAR) and Southern Africa (SA) (Figure 6.6). This finding brought up the question of whether the geographic distribution of the early hominin record, particularly of the Australopithecines, is biased by selective sampling (Gowlett 2004). This question implies that the record is influenced by preservation in the rapidly aggrading volcano-sedimentary sequences of the EAR and the protection in caves in the SA. If this argument is proven correct, it may change the idea that Australopithecines evolved in areas of savanna and woodlands, near water, and near mountains, as is broadly accepted in the discourse of human evolution science.

However, no evidence to support the above argument beyond the Koro Koro finding exists. Arguably, one can propose that the extreme dry conditions and intense erosion in deserts or alternatively the extremely moist conditions in the wet tropics may affect preservation. The case of the extreme arid areas is difficult to assess, despite the case of Koro Koro, because of many problems with preservation and the lack of rapid sedimentation that characterizes the EAR. At the opposite end of the climatic spectrum are the humid tropics, with visibility and taphonomic issues but also poorly researched, despite the possibility that Pliocene hominins and certainly some forms of early *Homo* could subsist in forests (Mercader 2002). Although the strong argument for the advantages of savannas for bipedal hominins seems to suffice, the problem is that limited investigations have been carried out in areas outside the savanna, which in a way is an issue of selective sampling. One can also assume that during the early dry phases caused by the early glaciations, the forests opened up into savannas, allowing the incursion of early hominins in areas now covered with rain forest. However, this would be very difficult to prove given the difficult conditions for preservation in the wet tropics.

A counter-argument to the issue of the alleged selective sampling and poor preservation of hominin remains outside the EAR and SA areas can be proven by the existence of a number of Mio-Pliocene and early Pleistocene faunal localities across Africa, where there is no evidence of hominin remains or tools (see locations in Werdelin 2010). One typical example is the Laangebaanweg quarry in the Cape Region of South Africa, whose fauna is contemporaneous with the Australopithecines in the Transvaal (merely 2000 km away), and yet no evidence of Australopithecines are found in the bone beds of the quarry.

Asia (Figure 6.5). Each locality is in fact a constellation of fragments of what once was a broad area of intermittent occupation spanning several thousands of years. Besides the fact that most of the record may have been modified by geomorphic processes over the tens of thousands of years, it provides us only with microscopic views of a big process of migrations, adaptations, occupations, abandonment, reoccupation, and transformation by post-depositional processes. Then, the question is how careful should we be with interpretations at one locality, or section.

Take the example of the Druze Marsh, a typical case of a relatively well-preserved record, consisting of deeply stratified open site (Figure 6.7). Within this series of sections with materials spanning from late Acheulian to Neolithic, a single layer of occupation contains the earliest Mousterian occupation (Figure 6.8). Through relative dating the layer is dated to MIS 5c (peaking around 96 ka) or MIS 5a (peaking around 82 ka) depending on the errors of U-Th dates of correlative deposits (Cordova et al. 2011; Ames and Cordova 2015). The question that paleoanthropologists may ask is: is this an occupation layer associated with anatomically modern humans moving up along desert refugia towards the Levant or with local Neanderthals?

As it is well known, Levantine sites on the Mediterranean coast bear remains of both Neandertals and AMS (Bar-Yosef 2000), and the dates in such localities seem to indicate that it is possible that at the Druze Marsh one of the two, or both may have existed. One problem is that the identified lithic industries associated with both groups at the time are similar (Bar-Yosef 2000; Henry 2003), which without hominin fossil remains it would be difficult to attribute to a particular hominin group. Therefore, jumping to conclusions based on a single finding, which often plagues the record, is a serious mistake. Geoarchaeology, however, with its multi-scalar view and the idea of contextual levels of interpretation (Figure 3.1) should help to determine whether such type of site-based assumptions is valid.

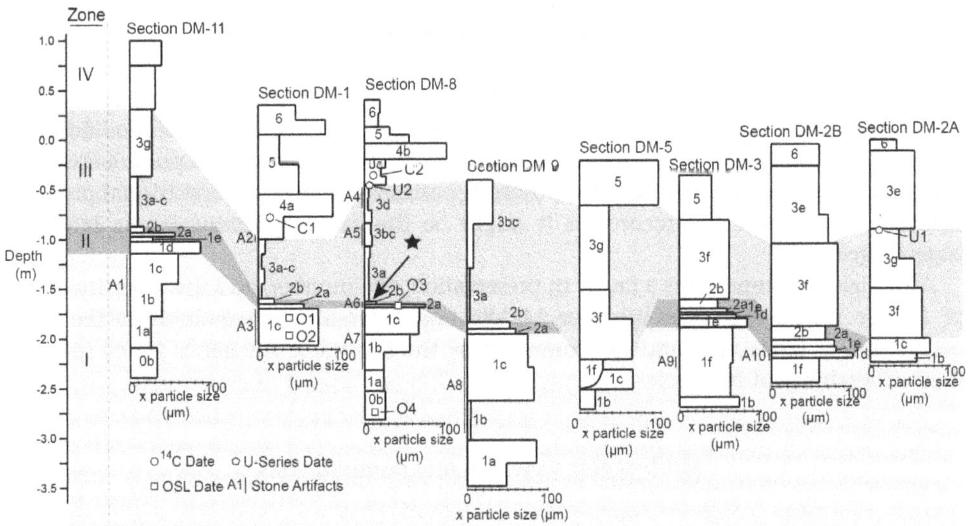

Figure 6.7. Sequences at the Druze Marsh in the Azraq Oasis, Jordan. Modified from Ames and Cordova (2015). The star indicates the stratigraphic location of the image in Figure 6.8.

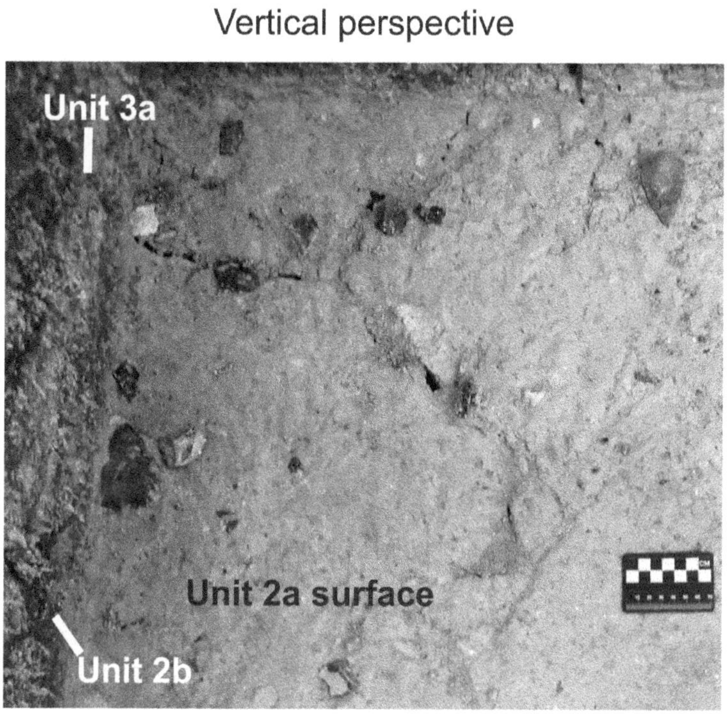

Figure 6.8. Vertical perspective of Mousterian lithics on an occupation floor in sedimentary layer 2b–3a, artifact zone A-6, in section DM-8 (see star in Figure 6.7 for stratigraphic position). Modified from Cordova et al. (2011).

CASE 6.1: CONTEXT AND SCALE IN THE OLDUVAI HOMININ RECORD

Geological stratigraphy, paleoenvironment, reconstruction of hominin behavioral-environmental patterns

The reconstruction of micro-environmental and behavioral aspects of human life in the archaeological record require fine scale views, obtainable in small areas and through micro-stratigraphic studies. However, this is a challenge due to the often dispersed (in time and space) and fragmented nature of the early hominin record in Africa. Nevertheless, obtaining aspects of hominin behavior from the record, such as defining activity areas and preferences for localities, has been tried in many paleoanthropological localities. The example discussed here shows some of the views on the particular case of activities and site choice in relation to local environments.

The Olduvai (Oldupai) Gorge is one of the most famous and heavily researched hominin site complexes in the eastern branch of the East African Rift (Figure 6.6). The deep cutting of the gorge exposed a series of volcanic, lacustrine, and alluvial layers (Figure 6.7) with abundant faunal assemblages. The hominin record of the greater Olduvai region includes localities such as the Laetoli and Peninj, is one of the richest and diverse hominin fossil

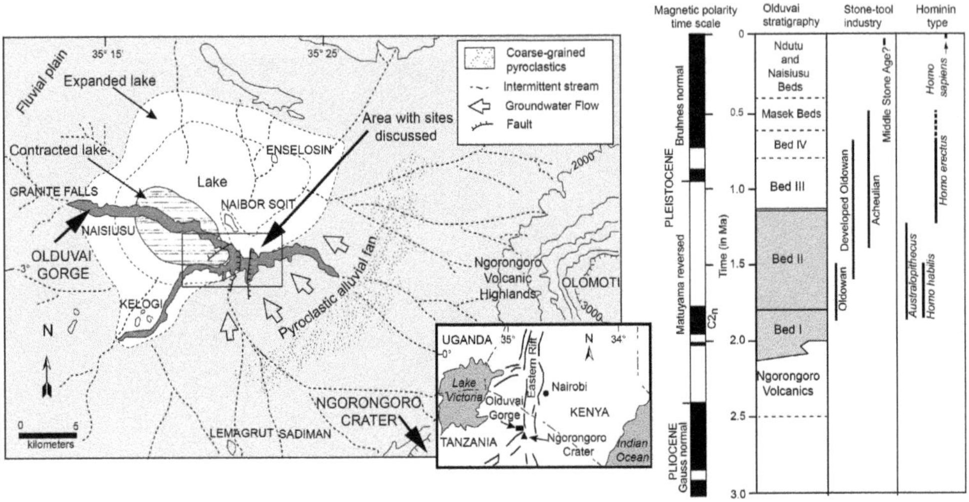

Figure 6.9. Olduvai area and general stratigraphic model of the Olduvai gorge sequence. Modified from Hay (1990) and Ashley et al. (2010).

regions of Africa as it contains almost seven species of the genera *Australopithecus*, *Paranthropus*, and *Homo* (Barboni 2014).

The nearby Ngorongoro volcanic massif also represents an important source of volcanic ash layers, which are useful for dating hominin occupations (Hay 1990; Ashley et al. 2010). The geological information obtained from the deposits themselves has been important in placing the fossil locations and correlative events in space and time. The sequence contains information on geological events of water impoundment into a lake or marsh interbedded with fluvial and pyroclastic layers (Figure 6.8), some of which are also tectonically deformed. Although tectonics played an important role in the hydrological changes in the former lacustrine basin, changes in the formation of the lakes and marshes occurred in combination with climatic pulses (Ashley et al. 2010; Maslin et al. 2014).

The discontinuous, but abundant, paleobotanical record has provided significant information for the reconstruction of local climatic and environmental changes, confirming some of the general patterns of research going from wooded environments to savannas. Interestingly, the effort in this reconstruction has been possible using analogs in the wooded and grassy ecosystems of the modern surroundings (Albert, Bamford, and Cabanes 2006; Albert 2015; Barboni 2014). Added to this paleobotanical record is the valuable faunal record (Cushing 2002), which is considerably abundant in the area, with numerous taphonomic studies involving experiments (Dominguez-Rodrigo, Barba, and Egeland 2007).

But despite the large time scale represented by the deposits, the question of the microstratigraphic focus on sediments associated with hominin fossils and tools has been of interest, particularly as a means to reconstruct behavioral aspects of the evolution from *Australopithecus* species into *Homo* species. The idea of reconstructing activity areas was proposed by Mary Leakey (1971) in terms of finding information in sediments, particularly what in Chapter 3 is referred to as traces (Table 3.2). Keeping in mind this proposal, the sediments associated with fossils and tools have been thoroughly studied.

Taphonomy studies on faunal remains have been the main source of information on activity areas (See Dominguez-Rodriguez 2007). However, the taphonomic record is not always sufficient, so it has to be complemented with reconstruction of vegetation and water resources in the surroundings, essentially the idea of ecological context. In some of the beds in the Olduvai sequence, this information came not only from the modified faunal remains, but also from the horizontal distribution of sedimentary facies (Ashley et al. 2010), paleolandscapes (Blumenschine et al. 2012), and the study of vegetation associated with burned areas and modified bone (Albert 2015).

Phytoliths helped verify the locations and the use of plants, particularly placing them in the context of the faunal remains and the sedimentary facies. The combination of proxies at discrete locations provided interesting results regarding human behavior, in this case of *Paranthropus boisei* and *Homo habilis*, but with the potential for similar studies on older hominins.

Some of the interesting findings of these micro-stratigraphy, multi-proxy studies show that hearths and modified bones are found in concentrations, which in the context of paleobotanical materials appear in areas with micro-remains of palm trees and other aquatic vegetation, suggesting a preference for protected areas close to water (Albert, Bamford, and Cabanes 2006). Likewise, the use of plants and animals at this location was significant, suggesting that early hominins, even before *Homo*, already practiced a noticeable use of their environment (Albert 2015).

The study then shows that micro-scale reconstructions are possible as long as the focus on obtaining information from traces is carried out from multiple disciplines. Although the reconstruction in this case is mainly carried out through other fields of environmental archaeology, geoarchaeology provided the context by adding information on the horizontal distribution of sedimentary facies, which in turn indicated the different hydrological features and micro-relief necessary to create the scene where hominin activities took place.

7

Geoarchaeology and the Anthropization of the World

INTRODUCTION

While hominization encompasses the evolutionary process of becoming human, *anthropization* refers to the appropriation by humans of unoccupied spaces. Anthropization occurred relatively late in the history of human evolution, thus becoming a characteristic of the genus *Homo*, and in particular of *Homo sapiens*. In this process, humans use these newly occupied spaces, previously part of the *pre-anthropic* world, incorporating them into the *anthropic* world.

Although the boundaries between the pre-anthropic and anthropic worlds are understood in principle, they are difficult to discern in practice, especially when their evidence in the record (archaeological, geoarchaeological, and paleoecological) is most of the time fuzzy, thus creating a temporal and spatial uncertainty that for a region, continent, or island is expressed in the question "When did humans first arrive?" In different terms this uncertainty is the time span when a human (or hominins) could or could have not been present in a determined region, continent, or island. Here such a temporal and spatial uncertainty is referred to as the *proto-anthropic* world.

To illustrate the important role of geoarchaeology in studying the anthropization process, this chapter refers primarily to the situations of the late Pleistocene peopling of Australia and the Americas. For that purpose, two cases are discussed in detail, one the issue of early sites with presumably modified megafaunal remains in North America (Case 7.1) and the dating and interpretation issues at the Lake Mungo site in southern Australia (Case 7.2).

THE PROTO-ANTHROPIC PERIOD AS A "GRAY ZONE"

The anthropization process across the world

The sequence of *Terrae* illustrates the process of anthropization across the world (Figure 6.3). But each Terra comprises long periods of time and large territories, reducing the possibility of understanding the proto-anthropic period in shorter time frames. Seen in a different way, the proto-anthropic uncertainty varies considerably across time and space (Figure 7.1)

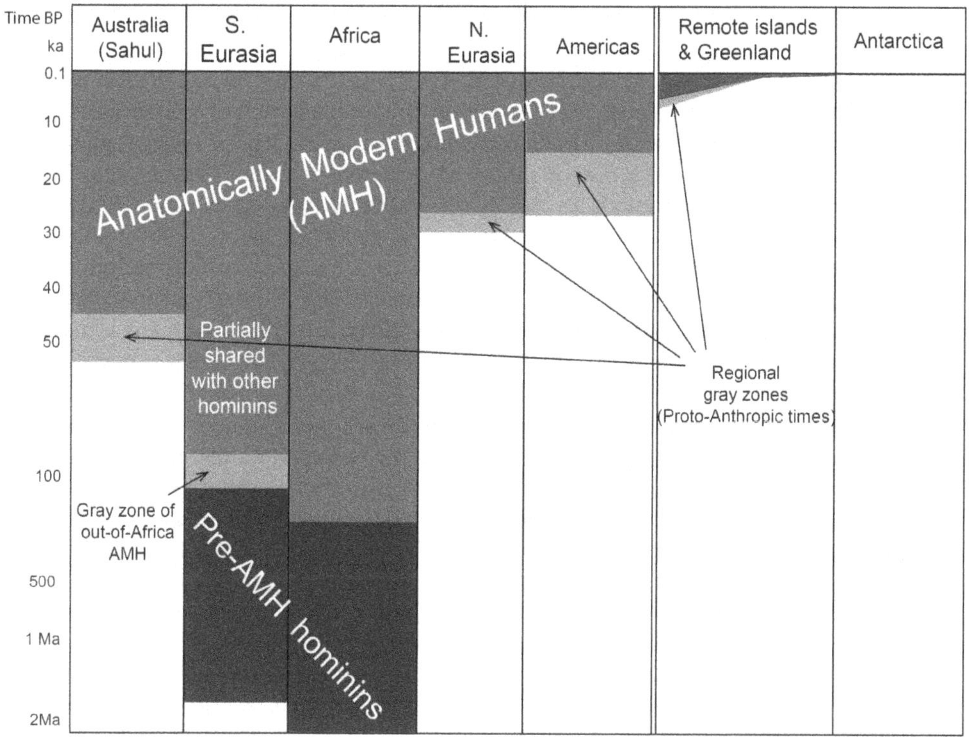

Figure 7.1. The global anthropization of the world across continents and regions.

The problem of defining the temporal and spatial dimensions of anthropization is first which hominin is used as a reference and which spaces is then considered. In essence, *Homo erectus* is the first one to allegedly have reached all parts of Africa and areas out of Africa. But often the focus of anthropization is on *Homo sapiens*, particularly its latest form, the anatomically modern human (AMH). Although archaic forms of *Homo sapiens* (e.g., Neanderthals and Denisovians) occupied areas of Eurasia, that were occupied earlier by *H. erectus*, perhaps more to the north.

The spread of anatomically modern humans (AMH) represents a wave moving faster, first into areas previously occupied by *Homo erectus* and archaic *Homo sapiens* (e.g., Neanderthals and Denisovians), and then into territories never occupied by any hominin ancestors. These territories include northern Eurasia, Australia, and the Americas, the polar continents (Greenland and Antarctica) and most of the insular world (Figure 7.1).

Not shown in the maps of Terrae (Figure 6.3) or in the broad time-frame of the proto-anthropic period are certain small islands in the Mediterranean that being surrounded by water even during the glacial sea-level drops were never reached by hominins until very late in the Pleistocene (ca. 13,000 BP), as is the notable case of Cyprus (Simmons 2012). This in turn raises the issue of humans being able to reach islands across open sea, which is a topic discussed for earlier colonization scenarios, namely the cases of Australia and the Americas.

GEOARCHAEOLOGY AND THE PRE- AND PROTO-ANTHROPIC PERIODS

The archaeological frontier dilemma

By definition, geoarchaeology operates in the field of archaeology, which creates a tie to the presence of humans, or at best hominins. Differently expressed, there is no archaeology if there is no evidence of past human activity, and there is no geoarchaeology if there is no archaeology. But in the context of the pre- and proto-anthropic world this dilemma becomes a problem, particularly for geoarchaeology, which with its geoscience side it is not bound by the absence of humans. In practical terms, this problem affects the ways the proto-anthropic is addressed in the broader field of archaeological science.

Understandably, the pre-anthropic world does not draw the attention and interest of archaeologists, and more specifically anthropological archaeologists, or in practical terms, archaeologists are not likely to take chances to explore the proto-anthropic world. However, oftentimes the discourses regarding early occupations involve assumptions on topics that pertain to the pre- and proto-anthropic world. To illustrate this situation, using the case of the early occupations in the Americas, the archaeological literature discusses the Pleistocene megafaunal extinctions in the context of a narrow sample of sites over a short period of time, usually during the short duration of the Clovis period (under 300 years). However, extinctions are processes that encompass aspects of biogeography that are not necessarily measured in short time frames and across small areas.

To put it differently, it is not possible to assume extinction processes, particularly those where humans are involved, if nothing is known about the environments and population dynamics of the extinct species. Thus, shouldn't we be looking at the pre- and proto-anthropic periods? How can we search for the "earliest" sites in the continent if we do not venture to explore the proto-anthropic period (i.e., the Gray Zone)?

It seems that although exploring the pre- and proto-anthropic periods may not be part of an archaeological agenda, but instead a multidisciplinary one, archaeology should seriously consider the importance of the pre- and proto-anthropic periods as a background to explaining ecological and social processes of early human activities in a region.

Geoarchaeological capabilities to study anthropic gray zones

In terms concurrent with the concept of geoarchaeological records described in Chapter 3, the spatial dimensions of the proto-anthropic periods across the world can be conceptualized in a series of objects that can be categorized into three main groups:

- Group 1: Quaternary deposits, soils, landforms, and biological remains with no prehistoric human remains (existing or reported) but correlative with dated prehistoric occupations in the same region.
- Group 2: Quaternary deposits, soils, and landforms with prehistoric human occupations that have not been yet recognized as archaeological sites.
- Group 3: Quaternary deposits, soils, and landforms with biological remains that do not fall within the two categories above, but there is suspicion that they may contain prehistoric human occupations or clues to them.

The main point to make here is that geoarchaeologists, regardless of their background, should be trained on studying one, two, or all the parts above. This means that like other disciplines outside archaeology (Figure 7.2) geoarchaeology plays a key role in investigating the pre- and proto-anthropic worlds with the purpose of investigating possibility of early human presence in the continent, and, if not, the landscapes and environments that preceded the arrival of humans.

It is important to understand that, even in cases of human colonization of new lands where dates of arrival are well established, the assessment environmental change caused by human arrivals requires the study of pre-anthropic landscapes and environments (Streeter et al. 2015). Therefore, geoarchaeology, along with other fields, particularly paleoecology, biogeography, and paleontology, should look in more depth to the proto-anthropic period, even if there is no apparent archaeology.

The discussion on the dimensions of the archaeological, geological, and geoarchaeological record, initiated in Chapter 3, comes full-circle when it comes to gray zone situations. While the geological record covers the entire time of the history of the planet, the geoarchaeological record covers only those segments of time where humans are present (anthropic world). Thus, the time limit of the geoarchaeological record should take a middle stance, or perhaps, as suggested in Figure 7.2, take over the proto-anthropic period, and perhaps even those processes occurring in the pre-anthropic period that have relevance in the formation of landscapes and environments that serve as the scene for anthropization.

Diversity of contexts and issues

Given the long time span of the proto-anthropic period (i.e., the gray zone) in both Australia and the Americas, the number and variety of early occupation sites and settings is diverse. Although sites occur in caves and rock shelters, most are open-air sites. In both regions, early sites are associated with alluvial, lacustrine, palustrine, eolian, cave and rock shelter, and spring deposits. In North America, several sites are found in loess deposits, since the Gray Zone in archaeology encompasses most of the Last Glacial Maximum (LGM) (22–18 ka).

Interestingly, in both continents, many are also associated with eolian sand deposits often associated with playas, as is the case of one of the earliest well-documented sites with human remains in Australia, Lake Mungo (Case 7.2) and many across the Great Plains of North America (Mandel 2008; Holliday 2009; Freeman 2016).

A recurrent problem at the contextual level of site in both Australia and the Americas is the taphonomic nature of the earliest human occupations, particularly when contamination or post-depositional disturbance (natural and human) affect the validity of numeric dates (Case 7.1).

In other situations, taphonomic problems may affect the possibilities of dating at all. Dating in many of the sites in Australia and the Americas has to do with the fragmentary nature of the record and the overall problem of palimpsest deposits. In other cases, dates are affected post-depositional transformation processes that have altered the archaeological context of the artifacts or faunal remains, sometimes placing them in secondary contexts. But given the number and variety of sites, it is difficult to generalize, requiring more of a case-by-case examination.

Another problem for the study of proto-anthropic sites in both Australia and the Americas is that the post-glacial sea level rise drowned many sites. This situation is even more detrimental for the Americas, where one of the likely routes of migration into the continent and spread to the south was along the Pacific coast (Anderson and Bisett 2015). But beyond this problem, the main issue is the enormous size of the continents, across which archaeological evidence is widespread and rarely visible.

Indirect evidence in the paleoecological record (e.g., pollen spectra and micro-charcoal record), which has been useful in cases such as human arrival in islands, is somewhat difficult to assess in large continents during the Pleistocene due to rapid climatic changes that led to unusual vegetation changes and unusual vegetation responses to fires. In some cases, time-transgressive phenomena and lack of high resolution may also be difficult to correlate with archaeological evidence.

Evidence of humans in more recently settled areas (Terrae 4–5) has actually survived changes not just because they occurred in more recent times, but because many of them occupy well-limited, small areas, such as islands. In most cases, such places were settled in post-Neolithic times, which meant a stronger impact on flora, fauna, and soils, which are more visible in records. Therefore, the big challenge of the pre- and proto-anthropic period lies mainly in the colonization of northern Eurasia, the Americas, and Australia.

CASE 7.1: BETWEEN PALEOONTOLOGICAL AND ARCHAEOLOGICAL SITES: GEOARCHAEOLOGICAL ISSUES IN PRE-CLOVIS MAMMOTH LOCALITIES

Background

The longstanding, but now defunct, Clovis-First paradigm is strongly based on the belief that the first arrival of humans in the American continent did not occur earlier than ca. 13,300 BP (around 11,000 ^{14}C), suggesting also that the passage occurred along an ice-free corridor between the Cordilleran and Continental ice sheets, thus permitting a migration between the Strait of Bearing and the unglaciated parts of North America (Waters and Stafford, Jr. 2013). But in the last two decades of the twentieth century, debates related to dates exist in a number of sites that pre-dated Clovis in North America, such as Cactus Hill, Meadowcroft Rock shelter (Pennsylvania), Paisley Caves (Oregon), Debra-Friedkins Site (Texas), Lagoa Santa (Brazil), and Monte Verde (Chile), among many others (Meltzer 2009; Freeman 2016).

Genetic studies along with the early presence of sites in South America have led to establishing the coastal route, as opposed to the previously accepted ice-free corridor in the interior, and many of the possible early sites of the coastal route are drowned by the post-glacial sea-level rise, which has prompted some underwater and coastal focus with little but promising results (Mackie et al. 2013; Anderson and Bissett 2015).

However, localities with putative human activities dating prior to the now more accepted Pre-Clovis (ca. cal. 15,000 BP) are appearing. Moreover, some pre-date the LGM (ca. cal. 22,000–18,000 BP), which even goes back the limit proposed by genetic studies. The archaeological community has reacted differently to these findings. However, whether the localities in discussion do represent early sporadic human presence in the continent, or cases where taphonomic problems make the evidence look "early," they represent a resource for geoarchaeology to study and for paleoecology to better understand ecosystems before human arrivals.

Paleoenvironments of proboscideans before their extinction

The two mammoth species known in North America towards the end of the Pleistocene are the woolly mammoth (*Mammuthus primigenius*), and the Columbia mammoth (*Mammuthus columbi*). The woolly mammoth occupied the areas closer to the ice sheets and the Arctic in what is known as the Mammoth Steppe (Figure 7.3a). The Columbia mammoth had a more widespread distribution, including overlaps with the range of the woolly mammoth and extensive areas in the central part of unglaciated North America as far as the southern part of the Mexican Altiplano Mexico (Figure 7.3b).

The extension of the Mammoth Steppe and its southern end, the CMS, may have varied through time, but the argument is that the period preceding the LGM (ca. 22,000–18,000) would have been the time of the optimal conditions for the Mammoth Steppe and the American Steppe (here the Columbia Mammoth Steppe), and a time when the Beringian Bridge and the northwestern corridor (ice-free corridor) would be open (Holen and Holen 2013). Alternatively, the sea level would have been lower and glaciers not yet fully extended to their maximum, opening up the possibility of a coastal route. Contextually, at the level of landscape and environment, this hypothesis makes sense, but at the level of setting and site, it faces many problems both in term of cultural and non-cultural processes.

Mammoths were not the only big mammals targeted by humans; other proboscideans were equally liked as prey by early Americans: the mastodon (*Mamut americanum*) and the gomphothere (*Cuveironious* sp.) (Haynes 1991). These species were found in areas with more trees (the mastodon) and tropical areas (the gomphothere). However, the bulk of potential archaeological sites of Pre-Clovis or early Paleoamerican age are largely those with mammoths.

Although not all Clovis and Pre-Clovis sites are found in association with proboscideans, in this study those with proboscidean remains, especially mammoths, are discussed because of their visibility. Unlike other sites, they tend to be more conspicuous when earth is dug out or stream cutbanks are exposed by erosion.

Pre-Clovis mammoth kill sites in North America

The discovery of mammoth killing and butchering in many of the Clovis sites in North America created an interest in the relation between proboscideans (mammoths, mastodons, and gomphotheres) and the earliest human inhabitants in the continent. These early findings defined subsistence of Clovis-time hunter-gatherers, suggesting the once-popular Overkill or Blitzkrieg. Accordingly, given the short frame of the Clovis period, in the order of 200 to 300 years (Waters and Stafford, Jr. 2007), it was evident that the killing happened quickly. However, as time passed, localities with apparently modified bones of proboscideans, among other megafauna, appeared with earlier dates, posing a tremendous challenge to the Overkill/Blitzkrieg Theory and overall to the Clovis-First paradigm.

Although many Pre-Clovis mammoth localities with evidence of human activity failed to convince the archaeological community, well-dated and highly convincing sites such as

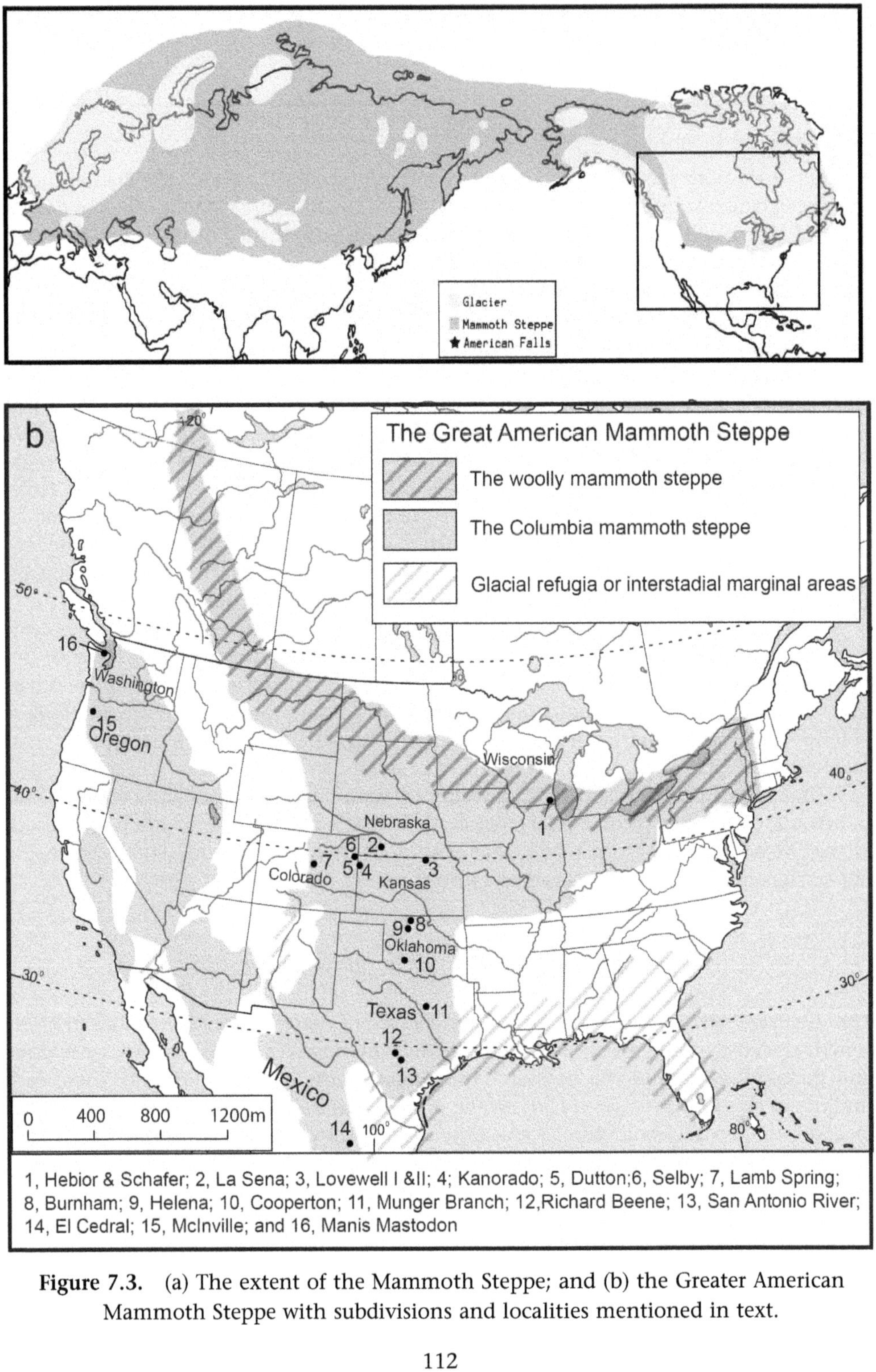

Figure 7.3. (a) The extent of the Mammoth Steppe; and (b) the Greater American Mammoth Steppe with subdivisions and localities mentioned in text.

the Schaefer and Hebior mammoths in southern Wisconsin (Overstreet and Kolb 2003), the Manis mastodon in Washington State (Waters et al. 2011), and the Kanorado assemblage in west Kansas (Mandel 2008), among other classic examples, suggest that humans not only arrived in the continent earlier than thought, but also that they were hunting the now-extinct megafauna.

The Cooperton Mammoth in Oklahoma, dated around 18,000 BP is an example, with the problem that the dates are questionable given the preservation of the bone (Holen and Holen 2011). Attempts to date bones in a museum collection have failed, and access to the original site, for the purpose of dating the layers, has been denied. The Cooperton mammoth had no clear association with lithic artifacts, except for a large rock purportedly used as a hammer to break the bones (Holen and Holen 2013).

Certainly, evidence of older mammoths with apparently modified bones in addition to the ones listed above exist. Many of the bases for suspecting human presence in early mammoths (usually those in the Gray Zone) is based on the breaking of the bones and the usual spiral breaking, or green-bone fracturing, and notches, which suggest activities for exploiting bones as possible tools or bone marrow (Holen and Holen 2011; 2013). However, the controversial aspect of green-bone breaking is a bit difficult to determine when certain conditions occurred at different stages during the formation of the record and conditions of the deposit (Haynes 2017).

Undoubtedly there are many depositional and post-depositional factors, natural and cultural, involved in each case of mammoth bones and possible artifacts in the record. However, the argument made here is a *geoarchaeological* one, that is, looking at the site taphonomy, particularly in the context of setting. This, obviously are related to the concept of landscape and environment, whose explanation is provided in large part by the distribution of the record in time and space.

Site taphonomy and setting

The alleged Pre-Clovis findings can be divided into two categories, one with dates no older than cal. 15,000 BP (Category 1), and another with dates older than 15,000 BP (Category 2) (Figure 7.3a). Seen in a different way, Category 1 sites are often within the range of the more accepted pre-Clovis sites, while Category 2 sites fall within the gray zone (Figure 7.2). The two categories can be analyzed in terms of convincing arguments, based on criteria at the contextual levels of site and setting (Table 7.1).

Based on the attributes in Table 7.1, the most convincing examples of pre-Clovis mammoth modification are the Schaefer Hebior sites (Figure 7.4), whose dates are concordant with the stratigraphy, particularly in the sense that little post-burial modification has occurred. The two sites are deeply buried with pedogenic development concordant with the time frame of its postglacial sedimentary and geomorphic position (i.e., the setting contextual level). Additionally, modification of the bones themselves is concordant with the use of tools present at the site (Overstreet and Kolb 2003). In summary, the checklist in Table 7.1 is complete. Likewise, the Manis mastodon in the state of Washington, with a flint point inserted in bone, is more convincing in the context of human presence (Waters et al. 2011), as it also fulfils most, if not all, of the attributes.

Two of the Category 2 sites, the La Sena and Lovewell sites in loess settings in Nebraska and Kansas, provide a good example of potential early American sites in the Gray Zone.

Their position with respect to the Peoria loess is consistent with their dates to place them during the LGM (Figure 7.6). The absence of lithics associated with the bone findings definitely raises suspicion that they may not be archaeological sites. But as Holen and Holen (2013) suggest, the animals were not butchered since the interest was only on procuring bone material and marrow. Similar settings occur in the cases of the Lamb Spring, Selby, and Dutton mammoth sites in northeastern Colorado, and the aforementioned Cooperton site in Oklahoma (see locations in Figure 7.3).

The Helena mammoth, excavated by Thomas Cox (2014), is located on a low terrace (Terrace 1) of a low-order tributary of the Salt-Fork of the Arkansas. The rib cage and

Table 7.1. Pieces of evidence that create a robust case for Pre-Clovis/Early American mammoth sites in North America

Attributes	Counter-attributes
At site contextual level	
Robust set of varied dates (more than one method) from undisturbed place	Few dates, particularly using one method may be equivocal and weak evidence
Strong evidence of bone modification – unequivocal difference of bone breakage	Bone modification that could be caused by natural causes (animals or geomorphic processes)
Artifacts in association with the carcass in the context of no post-depositional alteration	Artifacts are not clearly in association with the carcass, particularly in high-energy deposits or close to the surface
Artifacts consistent with the modification of the carcass	The artifacts are consistent with the butchering (bone marks) and breaking of bones, if not also with the killing of the animal
Layout of bones characteristic of human modification	Bones laid out under conditions of pedogenic or sedimentary post-depositional change, or by large animals
At the setting contextual level	
Relatively sealed deposit, preferably deeply buried	Unprotected, near-surface deposit may be affected by plowing, bone quarrying, erosion, and weathering
Original deposit – primary context	Post-depositional modifications may create contamination or disturbance of findings
Preferably found in low-energy fluvial, eolian, lacustrine, or paludal deposit	Colluvial or high-energy fluvial may be difficult to date, and are indication of post-burial alteration of remains

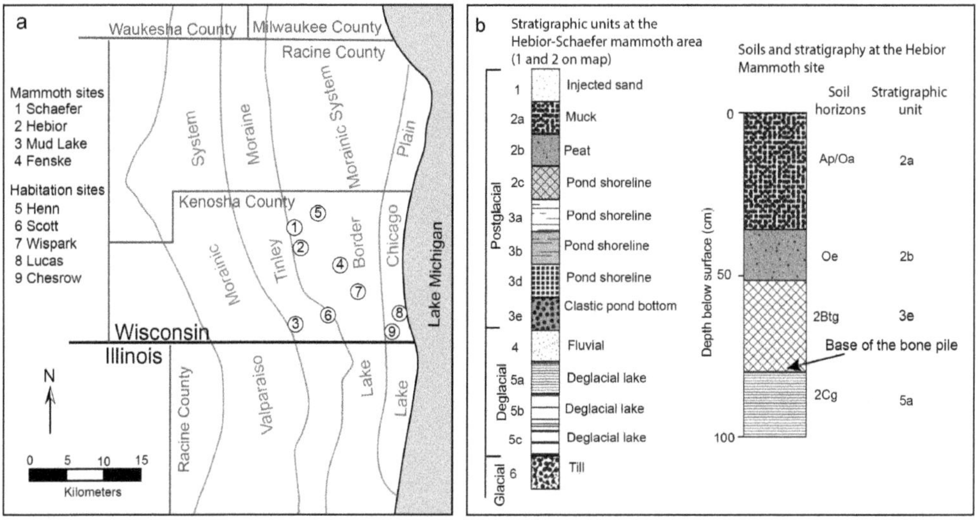

Figure 7.4. Schaefer and Hebior mammoth sites, among others; (a) geomorphological context and (b) stratigraphy. Modified from Overstreet and Kolb (2003).

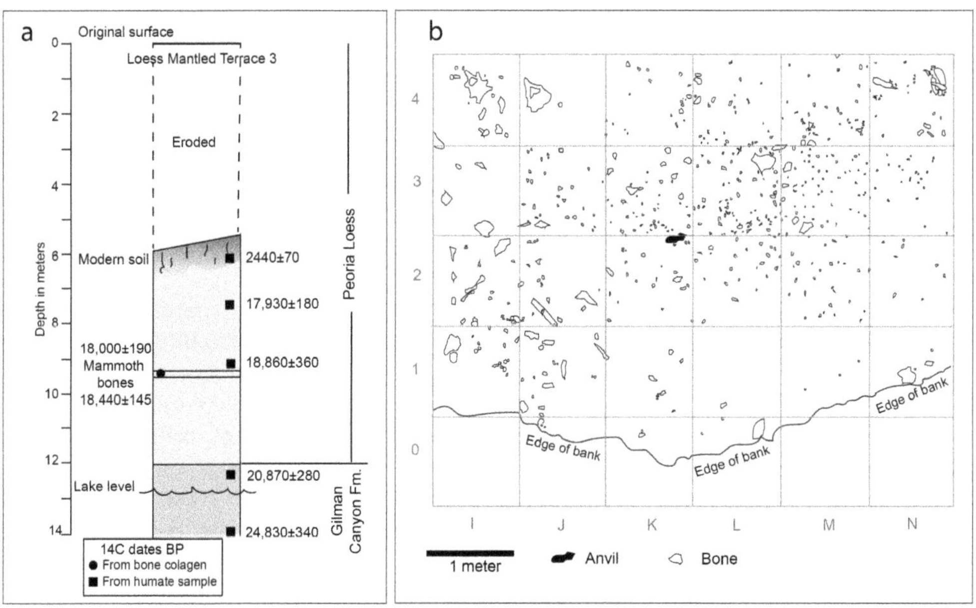

Figure 7.5. La Sena Mammoth site; a) stratigraphy, after Holen and May (2002); and b) mammoth remains, after Holen and Holen (2011).

Figure 7.6. Sketch of the stratigraphy at the Helena Mammoth site indicating approximate areas of mammoth findings. Based on Cox (2014).

vertebrae bones are lying directly in the transition from laminated alluvial silt-loam deposits, while the skull and the large limb bones are lying at the level of a colluvium deposit and a Bk horizon (Figure 7.6). The colluvial sediment originates from the upper terrace (Terrace 2). The finding suggests the dismembering of the animal bones, their breaking and placing in a higher level including one in the locations of the skull (Figure 7.6). In terms of re-arrangement of the bones and breaking, the findings bear similarities to the La Sena and Lovewell (Steve Holen, personal communication).

OSL dates at the Helena locality are considerably older than expected for an early Paleoamerican proto-anthropic site, in the order of 40,000 and 51,000 BP. The lower date comes from a low-energy laminated deposit consisting of silt and fine sand below the carcass. The upper date comes from the colluvial deposit engulfing the skull, which in turn raises some doubts about its validity given the mixed nature of colluvial material, originating from Terrace 1 (Figure 7.6).

The McMinnville Mammoth, Oregon (Bonnichsen, Full, and Reken 2006) which presents similar problems of bone modification as the Helena mammoth, also has relatively old dates (ca. 46,000). Numerous reports of localities with apparently modified mammoth remains but without associated artifacts have been reported in Texas (Thoms et al. 2007), as is the case of the Munger Branch Site (location in Figure 7.3). These examples show that either a natural process caused the bone breaking or the dates are affected by site taphonomy.

Post-depositional bone quarrying may suggest also the arrangement of the Helena limb bones in relation to other parts of the mammoth's body (Figure 7.6).

Interestingly, however, none of these sites (Helena, McInville, and Munger Branch) are not apparently associate with any lithic material. The Burnham site, located not far from the Helena site, has produced dates older than 30,000 (Wyckoff et al. 2003), which are concordant with the presence of *Bison latrifons*, but also with artifacts. However, the site has some taphonomic problems, which represents a contextual problem for the artifacts found in association with the bones, which are possibly in a disturbed context (Hill, Jr. 2005).

Lessons, alternative hypotheses, and potential studies

The cases of Category 1 mammoths discussed above are a matter not only of good documentation of the record, but also the luck of having an intact finding, that is, a record with little or no post-depositional modification. The case of Category 2 mammoth sites, which fail to fulfill one or more of the characteristics in Table 7.1, should not be considered cold cases, but examples that can help modify our methods to approach the problem of Early American/Pre-Clovis sites in a more substantial way. Among the various methodological and technical possibilities, the study of Category 2 sites can be tested with alternative and multiple hypotheses, as long as time and circumstances permit. In some cases, as in most CRM work, the possibilities of multiple testing are limited.

Category 2 sites, particularly the cases of La Sena, Lovewell, La Grande, and Cooperton, cannot be rejected a priori as Early American/Pre-Clovis sites. While there is no strong evidence to prove that such sites are archaeological sites, there is also no strong evidence to prove the contrary. Finally, if many of those localities with proboscidean or other megafaunal remains are studied with more detail, then a great contribution will be made to our understanding the environments preceding the peopling of the Americas, particularly with respect to the populations of animals that went extinct as human populations became more prominent in the landscape.

CASE 7.2: THE GEOARCHAEOLOGY OF EARLY AUSTRALIAN HUMAN ENVIRONMENTS AT LAKE MUNGO

The early Australian human occupation: The broad continental problem

Like its American counterpart, the earliest human occupation of Australia remains in the Gray Zone, with at least three issues still to be solved. First, the modes of human migration, possibly involving some sort of rafting from the nearby islands, remains unsolved (Webb 2006). Second, the interior migration south, which could have occurred through inland routes, or along the now-underwater coast, is still not clear (Hiscock 2013; Dennell et al. 2014). Third, problems of erosion and re-deposition, and the overall formation of palimpsests are issues that affect the preservation of the record and prevent the proper dating and placing of sites in the whole scheme of migration (Ward and Larcombe 2003; Head 2008). All together these issues have to be seen on the context of an enormous continent with a different configuration due to low sea levels (Figure 7.7), in the context of events that occurred at least 50,000 years ago, in a highly dynamic environment.

Ironically despite all the issues mentioned above, one of the relatively well-preserved sites, with a rich number of artifacts, faunal remains, and even human remains, is located in the southern part of the continent: the Lake Mungo site complex in the Willandra Lakes region in the arid interior of southwestern Australia (Figure 7.7). This site, despite many preservation problems, dating provides an insight into the local landscapes and environments during the early years of colonization. The information on the environment includes not only proxies for climate reconstruction, but also the hunting of fauna and evidence of human impact that led to the extinctions of the Pleistocene Australian fauna. Because of its importance as the earliest proven evidence of human occupation, marking an

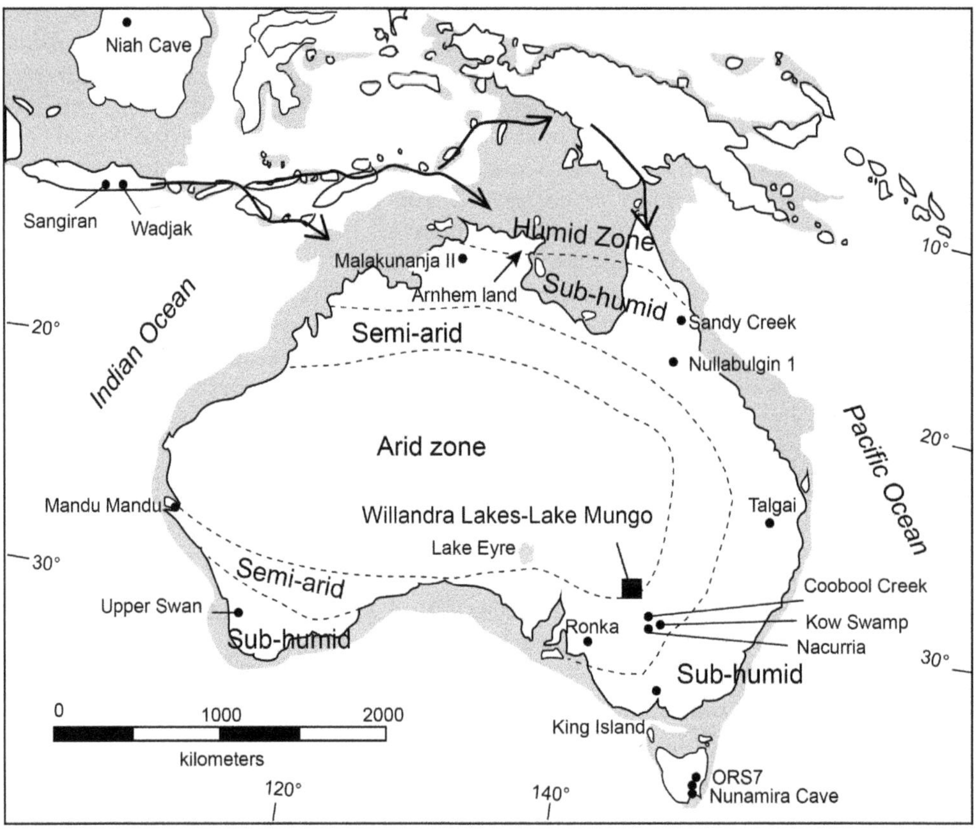

Figure 7.7. The Australian continent and adjacent lands with sites mentioned in text, with the lowest sea-level drop assumed for the LGM (ca. 126 m) and for MIS-3 (ca. 75 m), and possible routes into the continent. Sources of information: Webb (2003), Brown (1997), and Bowler et al. (2003).

Figure 7.8. Lake Mungo site: a) perspective with the dry lake bed in the foreground and the lunette (locally called Wall of China) in the background; b) erosion remnants of the Mungo unit resting on the Golgol unit; and c) erosion remnant of sandy-clay deposits of the Mungo layer.

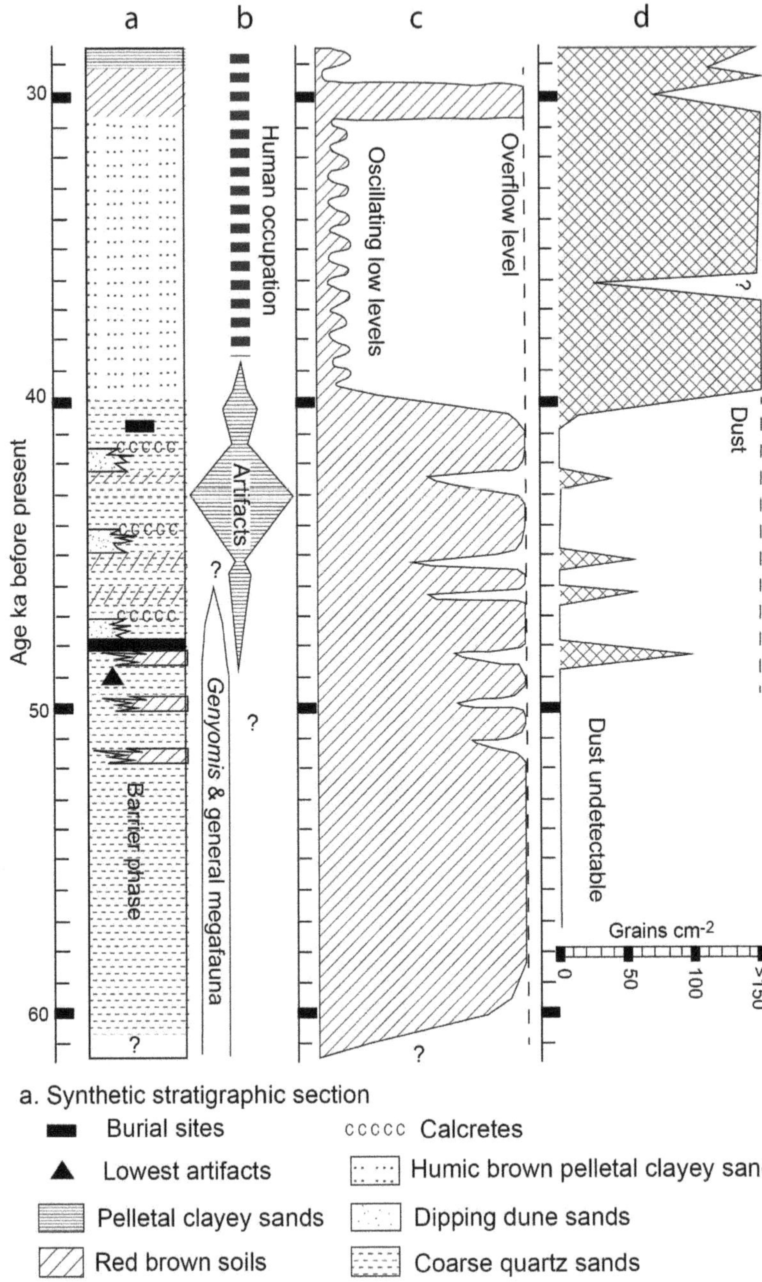

b. Human occupation, artifact frequency, and extinct megafauna

c. Lake levels

d. Desert dust accession from microscopic analysis

Figure 7.9. Stratigraphic, paleontological, cultural, and environmental relations produced from the record in Lake Mungo. Redrawn from Bowler et al. (2003).

end to the Australian gray zone of the proto-anthropic period, the Lake Mungo site complex deserves some discussion as a case in this chapter.

Lake Mungo as a microcosm of early Australian occupation

The site is located in the area known as the Willandra Lakes in the interior of the southern part of New South Wales between the rivers Darling and Murray. The lakes are a series of playa-type basins sometimes flooded by the rivers (Bowler 1998). The setting of the site is the shore of an ephemeral lake, or playa, embedded in a series of eolian sediments forming a lunette (Bowler 1998) (Figures 7.8b–c; 7.8a). Erosion has exposed the layers, particularly after sheep grazing altered the environment (Figure 7.8b–c).

The Mungo layer, formed by sands and sandy-clay deposits with shells (Figure 7.9a) contains the oldest human evidence, and a series of occupations, evident not only in human remains but also human traces (footprints), tools, hearths, and modified faunal remains. The archaeological record is punctuated, marking generally the times when climatic conditions made the area more habitable, particularly when the lake basin was filled and eolian accumulation waned, creating a stability that is marked by paleosols (Figure 7.8b). The oldest of these stable periods contain the oldest fossil, referred to as the Mungo lady (Mungo I), which is in the lower part of the Lower Mungo Unit, and the remains of a man (Mungo III) at the top of the same unit (Figure 7.2). Both findings fall within similar dates, centered around 42 ka (see Bowler et al. 2003).

The main archaeological issue faced at Lake Mungo has been the dating of the two oldest human remains, the Mungo lady and the Mungo man. It has been one of the most challenging issues, particularly because the two remains lie on the very edge of the time range of radiocarbon dating (ca. 50,000). However, as other methods were developed later in the century, dates began to emerge: OSL U-Th, and ESR (Thorne et al. 1999; Bowler and Magee 2000). Although the results using different dating techniques and localities vary and present some technical problems, all together they provide a time frame for the earliest occupation (Gillespie and Roberts 2000). The oldest dates using OSL on sediments and ESR on tooth enamel suggest burial age as old as 62.1 ka (Thorne et al. 1999). Although the human remains show a range depending on the method, the oldest dates (e.g., Thorne et al. 1999). This overlaps with some of the oldest evidence in the Lake Eyre region (Webb 2013), and the more studied sites of the Arnhem Land, Malkunja II, and Nawalabila (Figure 7.7), which suggest that Sahul (the continent encompassing the now disconnected lands of Australia, Tasmania, and New Guinea) could have been colonized before 60 ka.

Although not the only early Australian locality, data obtained from the Lake Mungo site complex has been important to reconstruct the earliest human-environmental interactions in Australia (Figure 7.9). Such interactions include climatic fluctuations and hunting of fauna, some of which went extinct (Bowler et al. 2003, Gillespie, Brook, and Baynes 2006). Thus, until another site that matches that richness of information, Lake Mungo will remain the microcosm of early colonization of Australia.

8

The Geoarchaeology of Hunter-Gatherer Landscapes

INTRODUCTION

The study of hunter-gatherers has created its own niche in archaeology because of specific questions pertinent to the unique behavioral aspects of foraging societies and the particular characteristics of their record (Prentiss 2014). Similarly, the geoarchaeological research of hunter-gatherer sites, settings, and landscapes has its own array of questions, ranging from methodology to interpretation of the record (e.g., Holdaway and Fanning 2014). However, these issues vary depending on the environment, landscape, time frame, and the degree of detail (scale) within which the geoarchaeological record of hunter gatherers is approached.

Geoarchaeological research of hunter-gatherer contexts shares similar problems with those of hominin sites (Chapter 6), although in most cases hunter-gatherers are seen as forager economies of anatomically modern humans. In this respect, most hunter-gatherer studies in archaeology and geoarchaeology pertain to anthropology and not to paleoanthropology.

The literature on the geoarchaeology of hunter-gatherer contexts is abundant, as are the number of specific topics and research examples. Therefore, the objective of this chapter is only to review the most important geoarchaeological approaches, particularly those at the contextual levels of the landscape and environmental change. For that purpose, two examples have been included, one from the Near East (Case 8.1) and one from the Great Plains of North America (Case 8.2)

HUNTER-GATHERER SOCIETIES AND THEIR ENVIRONMENTAL CONTEXTS

Defining hunter-gatherers

Broadly defined, hunter-gatherers are nomadic groups that obtain their food through foraging, a term that includes activities such as hunting, fishing, and collecting wild plant foods (Grove 2014). However, a broad spectrum of behavioral forms can be encountered among hunter-gatherers in the record, including those who also practice some form of horticulture, as with some groups in the Late Archaic in North America (Kay 1998), and those who own livestock, as some groups in the Kalahari (see Sadr 1997).

The different behavioral forms have led to typologies of hunter-gatherer groups, most of which have been based on the way they appropriate resources and their mobility patterns. Thus one can find terms such as foragers and collectors (i.e., specialized hunter-gatherers), generalists and specialists, and many others that combine many strategies into complex relations with the environment (Grove 2014; Prentiss 2014).

Because the archaeology of hunter-gatherers has traditionally relied on ethnographic research of modern hunter-gatherer groups, numerous debates have taken place regarding the interpretation of behavior in past foragers and their relations with agricultural societies (Yellen 1977; Wobst 1978; Sadr 1997), among many other debates.

The geographic aspects of hunter-gatherer societies

Despite the broad distribution of hunter-gatherer groups in different parts of the world at one time or another, the archaeological literature on the topic shows a geographic disproportion. There is considerably more research in semi-arid and arid environments, grasslands, and savannas and much less research in forested areas, particularly in the wet tropics.

In the dry areas of the Middle East, Africa, Australia, and western North America, surveys struggle to find sites with good integrity given the dynamic nature of these environments, normally because of the geomorphic dynamics of the slopes in those areas, which are conducive to destruction of the surface record through frequent runoff processes and/or deflation. Deeply buried records are possible, but normally in extremely arid areas the torrential nature of stream flow can be highly destructive. Alternatively, rock shelters, where present, tend to provide valuable information, because the record is likely to be more protected from erosion. Localities with more recurrent occupation (as is the case of Kharaneh IV in Case 8.1), the rapid accumulation of recent cultural materials protect the older ones.

In humid areas, particularly in the tropics, the wet and warm conditions are conducive to poor preservation of organic materials, particularly when soils are acidic. In such environments, rock shelters and caves, if available, are better preserved. As shown in studies in the tropical rainforest of the Congo, organic materials can be well preserved in caves and rock shelters despite the wet conditions (Mercader et al. 2003). But for open sites, the problem is that the dense vegetation and lack of accessibility impede the surveying of sites, then creating enormous gaps in the reconstruction of hunter-gatherer site patterns in these regions.

The part that preservation and visibility play is sometimes overlooked when placing the record in space and time. Sedimentary and archaeological palimpsests, on one hand, prevent the separation of periods particularly in areas with little deposition or deflation (see Chapter 4). On the other hand, the problem of preservation affects the appreciation of the hunter-gatherer record in time. For example, for very long, it was believed that population levels and complexity in Australia were much higher in the Holocene than in the Pleistocene, an idea that has in recent years been challenged through taphonomic studies over large territories (Holdaway, Fanning, and Rhodes 2008). Evidently, the more recent the record, the better preserved and the more visible (Fanning et al. 2009).

Another aspect that seems to be important to consider in the geography of hunter-gatherer studies is the influence of the more recent cultural landscapes. This issue affects the hunter-gatherer record of the past in two ways: one is through the destruction of the hunter-gatherer record, through the modification of the landscape by more recent societies, particularly those practicing sedentarism and agriculture, and the other is through the cultural bias archaeology has towards the more recent cultural periods.

In the Mesoamerican region, where most of the archaeology has been devoted to agricultural societies, locally known as Formative (Preclassic), Classic, and Postclassic, little or no attention has been given to previous hunter-gatherer archaeological periods (i.e., Paleoindian and Archaic). In Mexico, for example, hunter-gatherers of the pre-agricultural period are better known in areas of the north where monumental structures and urban settlements of the agricultural period are uncommon. But despite the bias towards more recent and monumental sites, some studies have stressed the importance of pre-agricultural groups, particularly those of the late Archaic in the development of the early Formative (agricultural) period (Flannery 1986; Niederberger-Betton 1987; Lesure 2011).

Other regions of the world with a long history of agricultural and urban settlement have produced substantial amounts of information on hunter-gatherer societies. The Levant, for example, has a great number of studies of hunter-gather economies preceding the development of agriculture, a matter that has always been seen as important for the development of agriculture itself (see Case 7.1). In part, the study of late hunter-gatherer societies (e.g., the Epipaleolithic or Mesolithic) have been of great interest in many parts of the Near East and Europe particularly because of their development into Neolithic societies. Thus, the idea is that late hunter-gatherer societies hold a key to understanding the transition to agriculture.

Studies of hunter-gatherer societies in the context of environmental change (i.e., physical and cultural change) are of great relevance for understanding the paths to Neolithization, whether locally developed or adopted by diffusion. The study of Late Stone Age hunting-gathering societies in arid and semi-arid southern Africa, along with the linguistics of modern groups, has been of great importance in understanding the processes involved in the adoption of pastoralism by hunter-gatherers (Sadr 1997).

Another aspect of geographic importance in the study of past hunter-gatherer societies is *mobility*, defined in simple terms as their daily, seasonal, and long-term movements across the landscape, an aspect that encapsulates the idea of nomadism. Mapping movements is one of the main objectives of archaeologists studying foraging groups, but it is also a difficult task if information on paleolandscapes does not exist. In many cases, though, assumptions are based on features and dynamics of the modern landscape. Although approaches using ethnographic examples have been used, mobility is something that apparently changes constantly based on climatic and, in general, environmental changes (Yellen 1978). Besides, the mobility of modern hunter-gatherers is most of the time limited by adjacent areas of farms, political boundaries, and other modern features (Yellen 1978; Grove 2014).

One of the approaches taken in environmental archaeology has been the use of stable isotopes on faunal remains, which help map certain migratory species that could drive the movements of their hunters. Examples, of this kind use stable isotopes on bison teeth and bones at kill sites in the North American Great Plains (e.g., Carlson et al. 2017). Although using isotopes is much easier given the record of the Paleoindian period, it is a problem for the Archaic period, where less material exists (see Case 8.2).

THE GEOARCHAEOLOGY OF HUNTER-GATHERER LANDSCAPES

The geographical aspect of hunting-gathering is one that geoarchaeology provides in many ways, not only through dating and site prospection and survey, but also through modeling based on studies of environmental change and mobility (e.g., isotopes). A broad regional view is as important as that of sites and settings, and one that provides many clues to the use of potential landscapes (Maher 2010) (see Case 8.1).

The environmental impact of hunter-gatherer groups

The general idea that hunter-gatherer groups live in harmony with nature and that they cause little damage to the environment has been challenged (e.g., Moran 2007). The concept of HMA (human means of adaptation) supports the idea that any type of subsistence activity impacts the environment (Dincauze 2000). The sole use of fire changes many aspects of vegetation, soil, faunal habitats, and in general the flows of energy in the ecosystem. In cases when hunter-gatherer groups move into previously unoccupied areas, extinctions do occur, presumably because of over-hunting and possibly through habitat destruction.

However, assessing the impact where hunter-gatherer groups have been established over a long period of time is difficult because, unlike the first arrivals, they have already managed the environment. In pollen records, sometimes the impact is difficult to assess given the noises imposed by climatic changes. Therefore, although in principle the impact on the landscape occurs through burning and overhunting, finding evidence in the record and linking it to hunter-gatherer groups is difficult. Nonetheless, coordinated research between archaeology, geoarchaeology, and paleoecology may in some areas provide some clues as to possible changes inflicted by major behavioral shifts among hunter-gatherers, as is the case of the relatively rapid changes occurred through the Epipaleolithic of the Levant (Case 8.1).

GEOARCHAEOLOGICAL APPROACHES TO HUNTER-GATHERER LANDSCAPES

Although the number of studies related to hunter-gatherer sites in geoarchaeology is vast, little has been written on the overall subject of the geoarchaeology of hunter-gatherer sites, except for regional studies (e.g., Holdaway and Fanning 2014) or *de rigueur* chapters on geoarchaeology in archaeological volumes focusing on surveys, or articles published in journals focusing on specific problems in the record. Although the literature is vast, there is no broad volume specifically on hunter-gatherer geoarchaeology.

Drawing from a number of publications, it is possible to signal some of the most important issues geoarchaeology tackles in the broad field of hunter-gatherer societies, past and present. These issues include aspects of preservation, visibility, context, and environment, all of which are briefly discussed in the rest of this section. In many cases these topics tend to be of regional importance, as in the widely distributed hunter-gatherer record in fluvial, lacustrine (playa), and eolian environments (e.g. Holliday 1997; Mayer 2002; Lovis, Arbogast, and Monaghan 2012; Kornfeld and Huckell 2016) as well as a large number of articles in journals. Likewise, geoarchaeological studies in every region of the world focus on particular problems defined by the specific geomorphic settings and environments.

Aspects of site formation processes are extremely important in studying the taphonomy of hunter-gathering sites, particularly where soil and sediments are conducive to the vertical movement of artifacts, creating problems of age determination (see Mayer 2002; Ward and Lacombe 2003). If not empirical then experimental taphonomic studies of are of great importance not only at the site level, but also at the regional level (landscape) for areas with geomorphic and pedogenic processes (Ward and Lacombe 2003; Holliday 2004; Goldberg and MacPhail 2006).

Off-site geoarchaeology might be a misnomer in most hunter-gatherer archaeological contexts, where the limits between sites and non-sites are difficult to establish due to faint site boundaries or modifications by post-depositional processes (per discussion in Chapter 3). However, off-site geoarchaeology in the sense of geoarchaeological tasks is enormously important in finding sediments for environmental reconstruction, and as means of reconnaissance for more, or better, preserved sites.

Areas that have been heavily settled in later times represent many problems, as the previous record tends to be destroyed by plowing or redeposition in soils and sediments in agricultural terraces. Thus, sometimes lithics of older periods appear in soils and structures of much later age. In such circumstances the best source of information on pre-agricultural sites exists in rapidly aggrading alluvial sediments. The problem here is that the agricultural period may also have created deep accumulations of sediments and incision (see Case 8.2).

Following the ethnographic and ethnoarchaeological component of hunter-gatherer research in archaeology, actualistic studies have been applied to the geoarchaeology of hunter-gatherer site formation under the rapidly growing field of enthnogeoarchaeology (Chapter 16) and the more conventional field of experimental geoarchaeology. However, although these two fields have become popular, the hunter-gatherer focus has lagged behind other cultural ecological areas.

CASE 8.1: EPIPALEOLITHIC HUNTER-GATHERERS OF THE EASTERN LEVANT: SITES, SETTINGS, AND LANDSCAPES IN A RAPIDLY CHANGING ENVIRONMENT

The idea that hunter-gatherer sites are mainly seasonal camps spread across the landscape is challenged by some cases where more permanent sites, even if seasonal, permitted accumulation of cultural materials to create more visible remains in the landscape. This is the case of the Epipaleolithic of the Levant where some sites even display more permanent structures and artifacts (e.g., grounding stones) that precede the domestication of plants and animals by several thousands of years.

The Epipaleolithic in the Southern Levant (Figure 8.1) spans a long period of time from the Last Glacial Maximum to the beginning of the Holocene, with the latter roughly coinciding with the first Neolithic settlements (Figure 8.2). This broad period of time encompassed profound climatic changes related to global dynamics concatenated with the deglaciation and cultural changes in large part led by climatic changes (Figure 8.2).

Amid the many cultural changes, the Late Epipaleolithic, i.e., Natufian period (Figure 8.2), represents the crucial steps in the evolution of human groups that led to the rise of social complexity, the emergence of sedentary village life, and the adoption of food production (Maher et al. 2012). But this change was neither linear nor clear-cut, because of their variability through the landscapes of the region and the rapid, contrasting changes during sub-periods encompassing over 10,000 years (Figure 8.2).

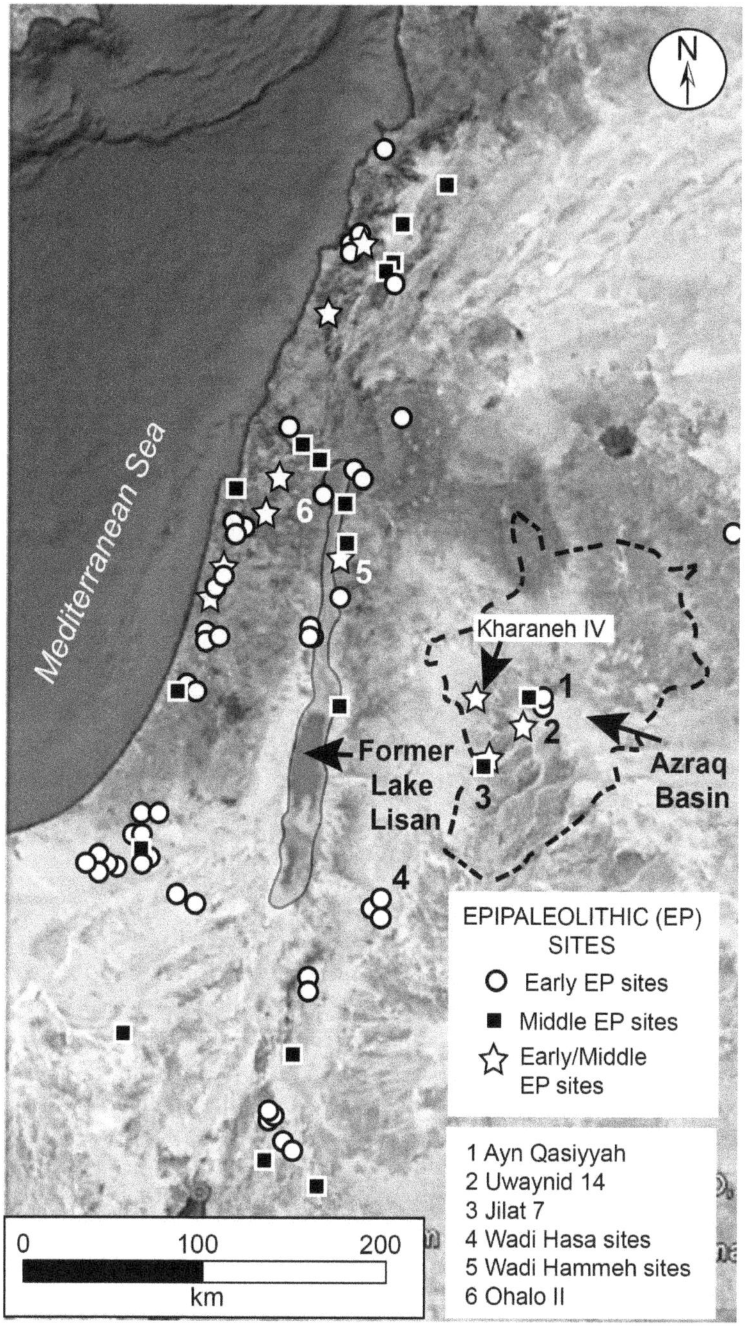

Figure 8.1. The Epipaleolithic of the Southern Levant. Modified from Maher, Richter, and Stock (2012).

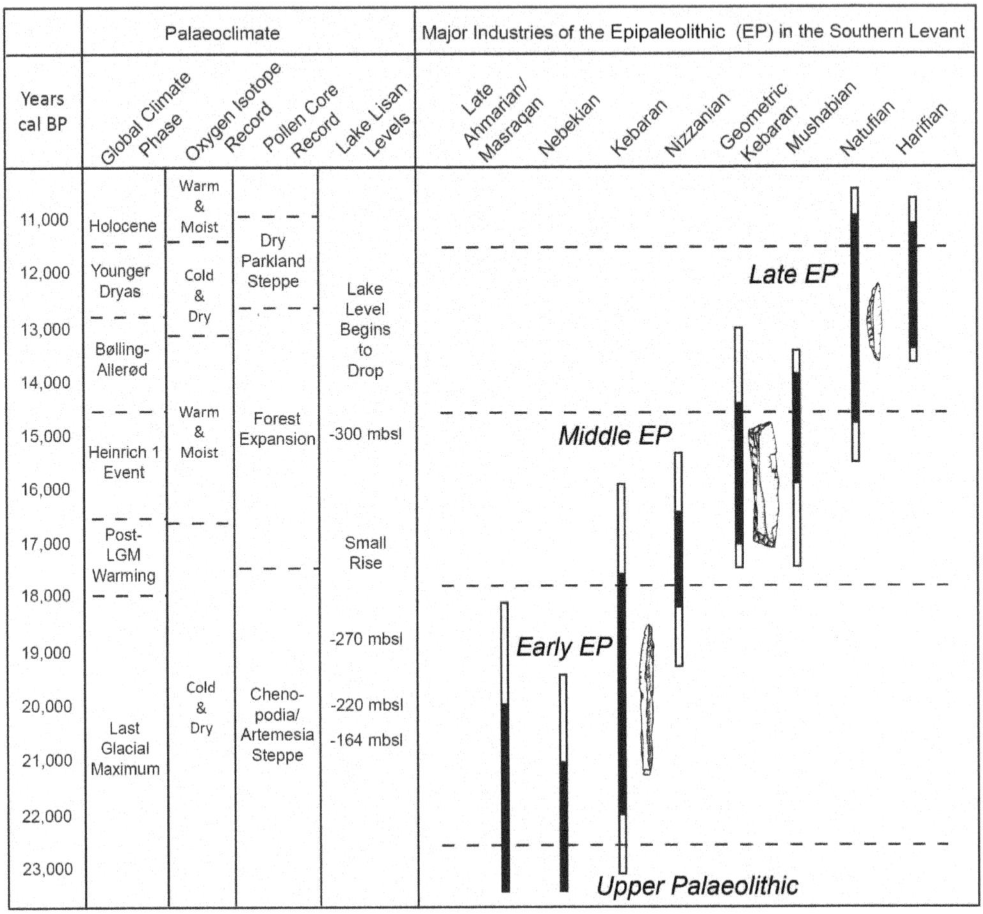

Figure 8.2. Cultural chronological scheme for the Epipaleolithic of the southern Levant alongside major climatic events. The bars for each phase represent calibrated radiocarbon dates (black) along with associated errors (white). Modified from Maher, Richter, and Stock (2012).

The challenges presented by climatic changes were characterized by a warming trend with alternation of warm and cold phases. The warm phases, presumably more conducive to moist conditions, were short and ended abruptly (Bar Matthews et al. 2003; Robinson, Sellwood, and Valdes 2006), followed by short cold periods characterized by atmospheric dryness, with the Younger Dryas being the last cold phase, coinciding with the Natufian period. The rapid climatic changes meant rapid changes in hydrology, flora, and fauna, which were the base of Epipaleolithic subsistence, which created insecurity, but at the same time communities learned from the experience to develop better adaptive strategies (Rosen and Rivera-Collazo 2012).

The role of geoarchaeology in the study of the Epipaleolithic has ranged from studies at the level of site, particularly understanding formation and transformation processes, at the level of setting and landscape, useful for locating sites and explaining the preferences for settlement, and at the level of environment, which encompasses aspects of climatic and environmental change and the cultural adaptation processes associated with them.

For many decades, most of the research was concentrated along the Mediterranean corridor, encompassing mainly the Mediterranean shrublands and steppe, and only recently directing attention to the more arid and hyperarid parts of the Levant, namely the Negev, the Jordan-Dead Sea Rift valley, and the Jordanian Desert.

In the Jordan Valley the case of Wadi Hammeh is a good example of semi-permanent locations near the shores of the shrinking lake but close to fresh-water sources from springs and streams (Macumber 2001). Similarly, the Wadi al-Hasa in the semi-arid Irano-Turanian steppe has been the subject of research particularly around paleolake Hasa (Figure 8.1). In southern Jordan, Epipaleolithic sites have been located mainly on stream terraces and around springs (Henry 1995).

The Azraq Basin lies in the arid and hyperarid parts of the Syro-Arabian Desert (Figure 8.1). The basin contains a number of Epipaleolithic sites in a variety of settings, along wadis, near springs and ephemeral lakes and swamps, or on basalt plateaus near springs and water pools (Maher 2010; Maher et al. 2012; Jones et al. 2016). Despite being in arid and hyperarid areas, the Azraq basin sites at these locations are by no means ephemeral and simple, but more permanent and stratigraphically complex (Jones et al. 2016; Maher et al. 2016).

The site of Kharaneh IV is located on the banks of a wadi, or ephemeral drainage (Figure 8.3). Geoarchaeological research determined that it is built on a wetland deposit, an occurrence in other sites in the Azraq Basin. Most of the wetland deposits have been eroded, but the cultural deposits on top of the wetland protected it from erosion (Jones et al. 2016). It is evident perhaps that the wetlands were widespread along the wadi, as occurred also in other neighboring wadis, as is the case of Al-Jilat (Figure 8.1). Thus, despite of its modern location on the banks of an ephemeral stream, the subsistence of the site at the time of occupation was based on a wetland environment.

The Kharaneh IV site can be described as an aggregation of occupations where recent excavations have exposed a complex way of life, not only in terms of varied foods, but also in terms of behavior (Maher et al. 2012). Although evidently hunting, gathering, and probably fishing in local and the nearby wetlands as the the base of the economy of its inhabitants, the site is far from being a simple seasonal camp.

The site is stratified, suggesting several occupations, and occupations that were more permanent than just a season, as structures of dwellings are apparent (Figure 8.4). In many ways, such structures are uncommon on pre-Neolithic site, and although some have been reported in the moister parts of the Levant, near the Mediterranean coast, the case seems unusual for an arid area. However, more than breaking standards created by a tiny record, Kharaneh IV, as well as other recently excavated sites in the Azraq Basin show how complex hunter-gatherer societies were even in the face of climatic challenges of the time.

Kharaneh IV, as well as many other sites in the southern Levant, show that the evolution of society in the Terminal Pleistocene was reaching a complexity that most likely was an important step to developing agriculture (Henry 1995). However, connection between semi-sedentary hunter-gatherer societies and agricultural ones is not as simple, as it requires plenty of data from the record across broad spatial and temporal scales. Geoarchaeology with its approaches, methods and techniques, is helping understand such complex relations between society and environment, even in landscapes where processes are conducive to destroying the record.

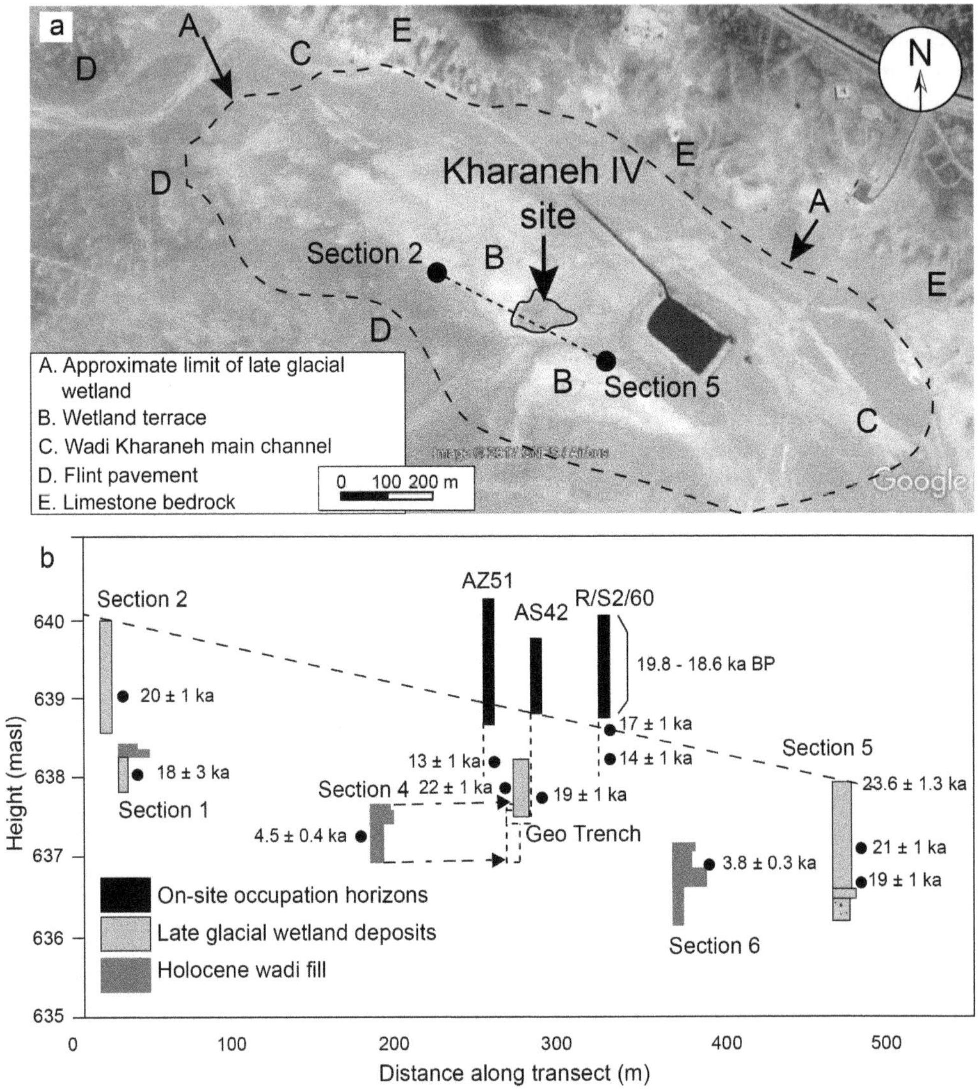

Figure 8.3. a) Google Earth image showing geomorphic of the Epipaleolithic site of Kharaneh IV, showing the relict geomorphic features of the landscape at the time of the site's occupation, and the location of the transect; b) sections along a transect showing dates for the different natural and cultural units. Based on Jones et al. (2016).

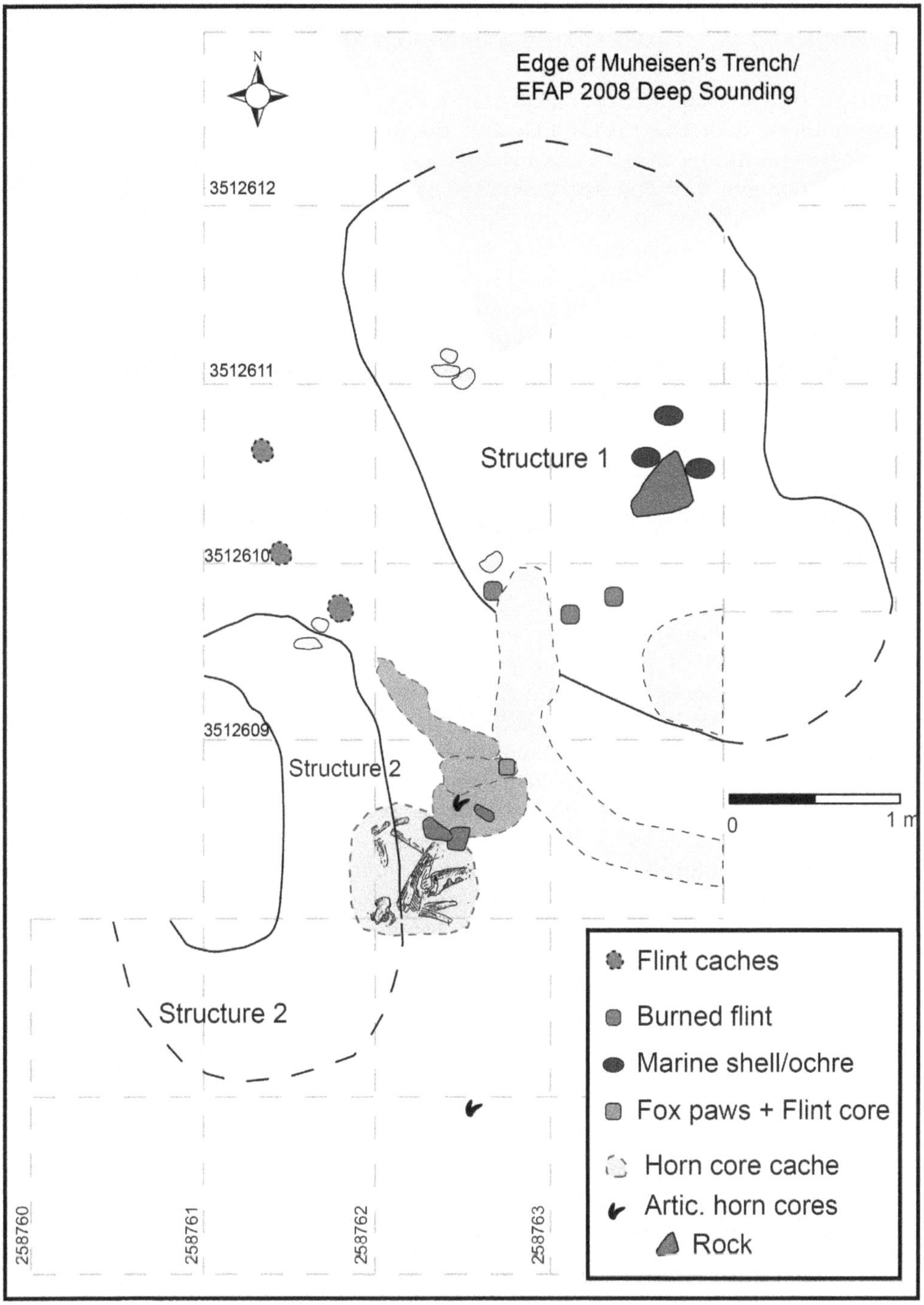

Figure 8.4. Archaeological map of an area of the site, showing the base of a dwelling structure. Modified from Maher et al. (2016).

CASE 8.2: THE GEOARCHAEOLOGY OF THE ARCHAIC PERIOD IN THE GREAT PLAINS OF NORTH AMERICA

In the North American archaeological chronology, the Archaic encompasses the time between the end of the Paleoindian and the appearance of farming communities. However, its beginning and end vary regionally across the continent. In the Great Plains, the Archaic extends from approximately 8500 BP to roughly sometime between 3000 and 2000 BP, depending on the first appearance of horticulture (Frison 1998; Kay 1998). Although the definition of the Archaic in time, space, and culture is still a matter of debate, its salient characteristic is that it is dominated by societies presumably depending on bison hunting and gathering. Nonetheless, its distinction from the Paleoindian, which is also part of the hunting-gathering period in North America, is a matter of debate among archaeologists, to the point that the term "Paleoarchaic" has become of use to include both the Paleoindian and the Archaic (Jones et al. 2003; Blackmarr and Hofman 2006).

But despite the difficulties with establishing the Archaic and its subdivisions (early, middle, and late) as archaeological periods, the hunter-gatherer societies in the Great Plains underwent several environmental changes during the Archaic, including a persistent period of high temperatures and reduced effective precipitation, referred to as the Hypsithermal (also known as the Altithermal) (Meltzer 1999; Mandel 2006).

Thus, the idea of a persistently dry Archaic began to be addressed, as the archaeological record in the Great Plains was scarce compared to the previous (i.e., Paleoindian) and later periods, despite its relatively long time span (between 5000 and 6000 years). In that respect, the prevailing hypotheses to explain such a scarcity of Archaic sites gravitated around the idea that the persistent low effective precipitation levels meant scarce resources (e.g., game, vegetation, and water sources), which either drove people out of the area or reduced their populations (Reeves 1973; Sheehan 1995; Meltzer 1999).

In the broad context of the record of the Archaic period, the early Archaic phase (roughly 5550–2500 BCE) in particular has the most obvious site scarcity, even in areas that have been well surveyed, suggesting a cultural response to an unfavorable climatic change (Sheehan 1995). However, this prevailing idea, which combined climatic, geologic, and cultural implications, was at the time difficult to assess because of the lack of data (Reeves 1973).

The Great Plains region possesses a sparse climatic record compared with other parts of the USA, which made it even more complicated to back the influence of a dry Hypsithermal on hunting-gathering groups. The matter is also complicated because this region encompasses a large area across a broad range of latitudes with a variety of climates and sources of precipitation and a high diversity of resources (Figure 8.5). Moreover, for such a large region, high-resolution climatic records from lakes for the mid-Holocene are sparse, particularly for the southern plains where the only lakes (playas) dry out during dry phases. Many of the paleoclimatic records, however, come either from fragmented sections, or from geomorphic sources, that is eolian and alluvial records.

The sparse climatic records, however, reveal not only variability of wet-dry cycles across the Great Plains, but also variability during the Hypsithermal (Hall 1988; Holliday 1989; Forman, Oglesby, and Webb 2001; Clark et al. 2002; Halfen and Johnson 2013). Thus, although predominantly dry, the Hypsithermal was characterized by extreme dry-wet fluctuations occurring at different times across the region.

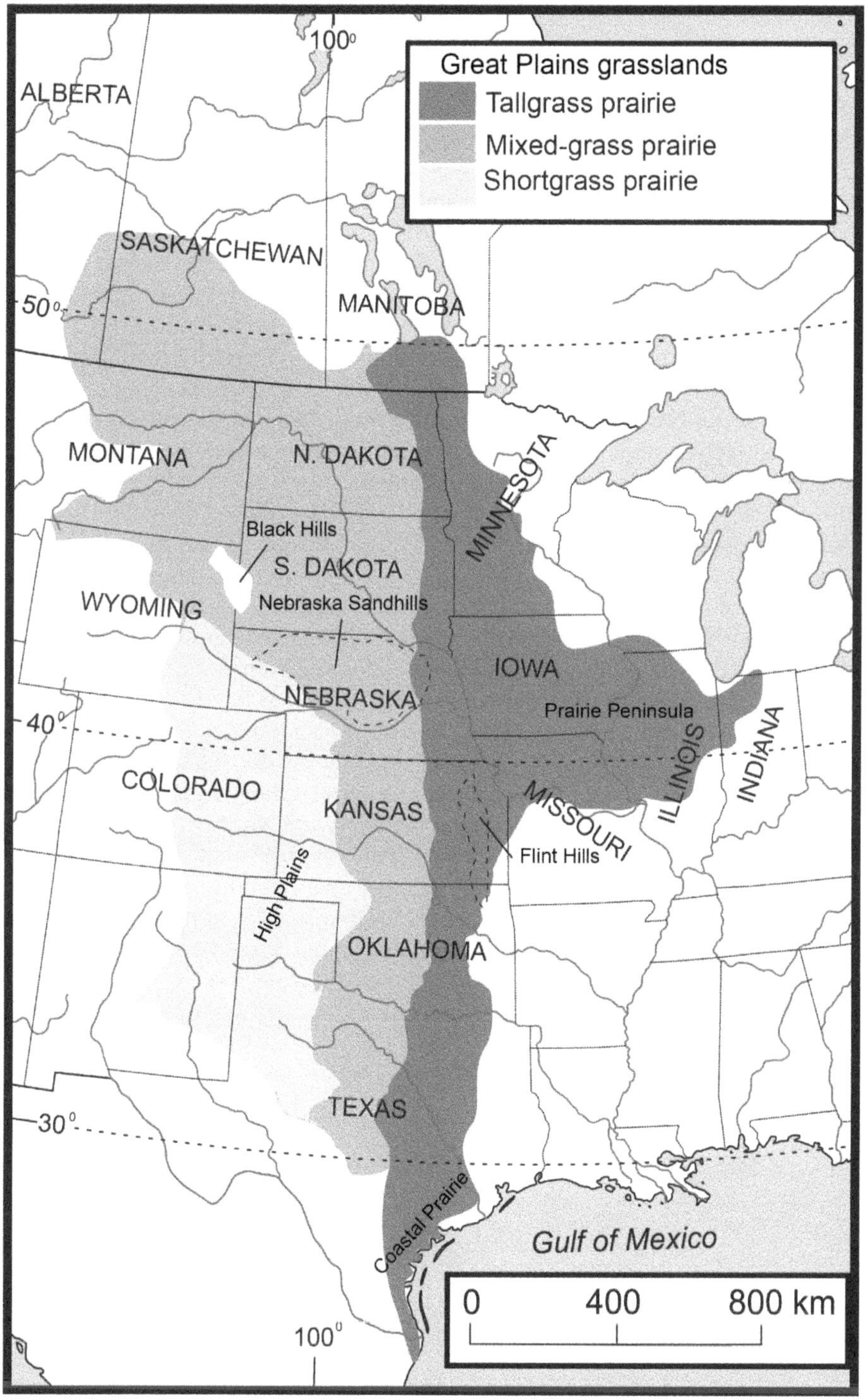

Figure 8.5. The modern eco-regions of the Great Plains grasslands.

The impact of dry climatic conditions has been seen in archaeology as an important determinant for the biomass of grasses and bison, the base of subsistence of hunter-gatherer groups in the Plains. The rapid diminution of the bison (i.e., reduction in body mass) during the early parts of the Hypsithermal, although thought to have been the result of predation, seems to be related to the poor environmental conditions that characterized the period (Hill, Hill, and Widga 2008).

After more studies became available, David Meltzer (1999) reassessed the issue in the context of newly studied sites and paleoenvironmental data suggesting that despite the dry conditions, population would have adapted, particularly because sites began to be reported in the driest part of the Great Plains, namely the southern High Plains. This questions the influence of climate, i.e., dryness, in the sparse archaeological record of the Archaic.

Although many studies confirmed the idea of a dry Hypsithermal, isotopic studies suggest periods with oscillations between dry and wet spells that may have caused changes and reappearance of bison in some areas (Lohse et al. 2014). In the end the idea of lack of settlement seems to be more related to movements from area to area depending on rapidly changing conditions and not the lack of settlement.

Still the problem of site visibility was still not completely solved, a matter that fell within the goals of geoarchaeologists working in the region. Thus, in addition to many papers published in the late 1980s and early 1990s, a collection of research addressing the geoarchaeology of the Archaic period across North America (Bettis III and Hajic 1995) began to shed light on the complex visibility issues of the archaeological record of that period mainly by looking at several localities across the plains and even along stream basins (Mandel 1995). The focus of research began to connect not only developments in interfluvial areas (e.g., playas and sand dune development), but also the temporal and spatial relations along different stream orders (Ferring 2001; Bettis III and Mandel 2002; Mandel 2006).

The first aspect that comes to light is the rapid changes occurred in the fluvial morphology during the Middle Holocene, which contrasts with the relative stability of the Paleoindian alluvial landscape (Mandel 2006). Thus the preservation of Paleoindian sites, many in high terraces, contrasts with the series of channel entrenchment and fill phases during the Archaic. The case of the Loup River in Nebraska (May and Holen 2014), although not studied for the purpose of explaining geomorphological developments during the Middle Holocene, epitomizes this interesting process (Figure 8.6).

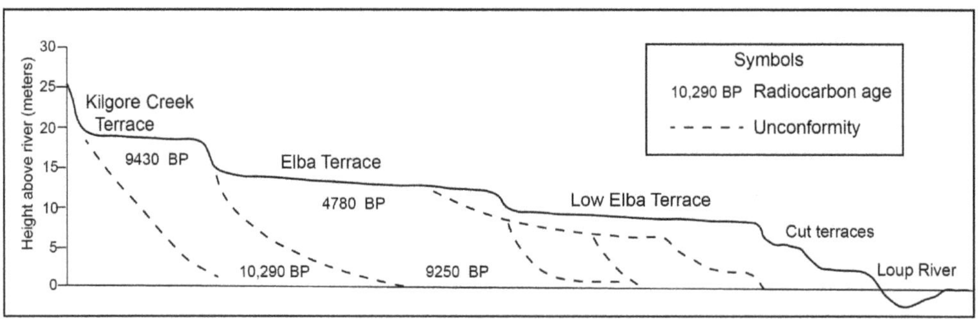

Figure 8.6. Generalized cross-section of Loup River geomorphic units and association with Paleoindian and Archaic sites. Re-drawn from May and Holen (2014).

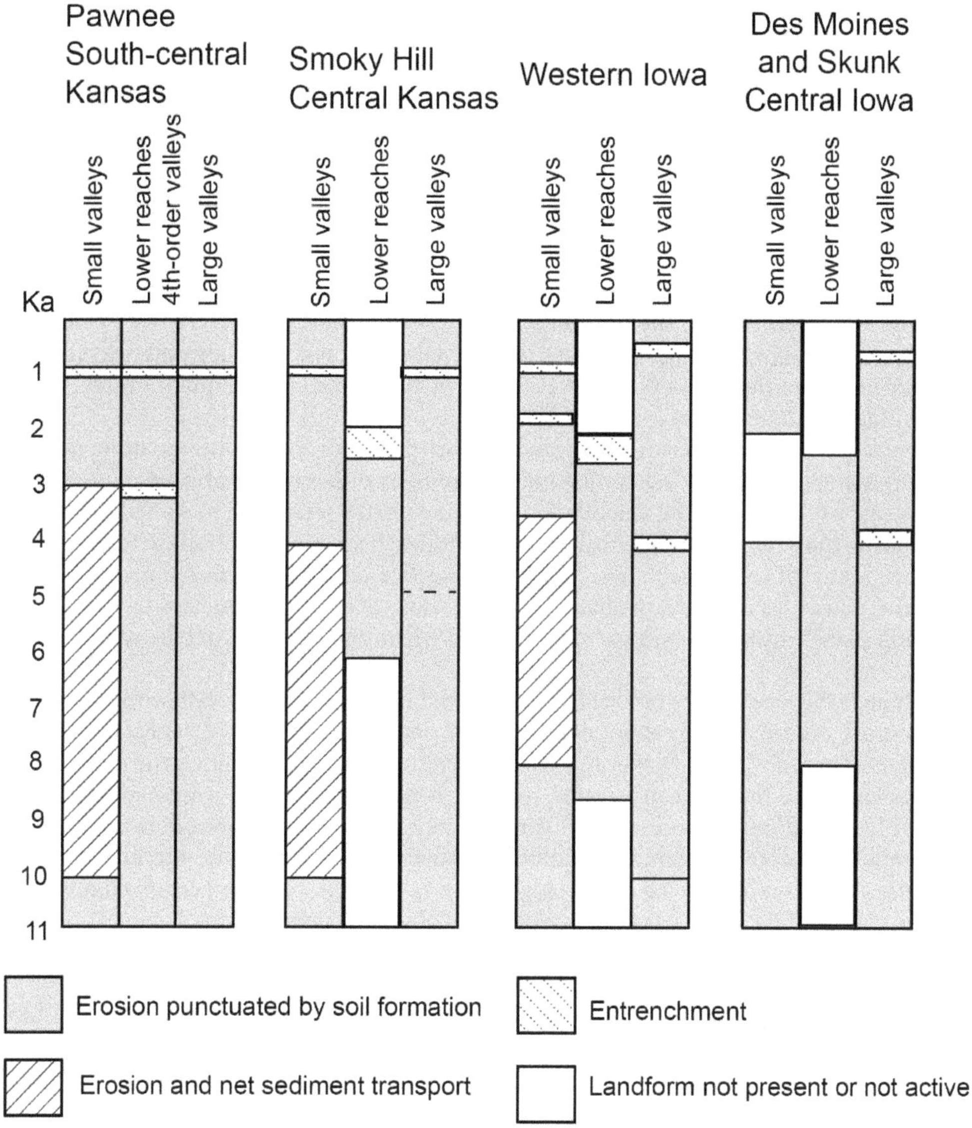

Figure 8.7. Summary of alluvial/erosion developments in different components of the drainage network at four localities in the central Great Plains (modified from Bettis III and Mandel 2002).

Among many, the example portrayed in Figure 8.6, exemplifies the case of scouring and erosion in a cross section, but not longitudinally along the basin and interfluvial surfaces. Thus, studies comparing sequences across and along fluvial catchments show that responses to climatic controls vary considerably in time and space between large (high-order) and small (low-order) streams and between stream and interfluvial areas (Figure 8.7).

The result of these changes is that most sites in slopes or in low-order tributaries ($<5^{th}$ order) eroded, where downstream, the along the high-order streams sites were deeply buried (Bettis III and Mandel 2002; Mandel 2006; Beeton and Mandel 2006; Murphy et al. 2014). Additionally, alluvial fans formed during the drier phases also burying site occupations under rapid sedimentation (Bettis III and Mandel 2002; Mandel 2006). This dynamic development in the fluvial systems of the Great Plains translates into the destruction of the record, particularly in interfluvial slopes and small streams, except when they are deeply buried and exposed sometimes in cut banks of subsequent entrenchment phases (Mandel 2006; 2008).

On flat interfluvial surfaces, as is common in the High Plains, the relation between geomorphological history and settlement is expressed differently, but more consistently. The playas, which during the Paleoindian contained more water and more sedimentation, were during the Archaic for the most part dry. Although seasonal water attracted game and humans, many of the remains were buried in lunette deposits where sediments deflated from the playas accumulate (Holliday 1989; Mandel 2006). The lunettes have been the common place where occupations occurred and where most of the Archaic record can be found.

Although the possibilities of Archaic site size and number may correlate with the scarcity of resources due to unfavorable environmental conditions, aspects of site sampling are important (Sheehan 1995). However, the case represented here exemplifies the idea that the cultural landscape does not necessarily reflect the complete history of human occupation because the geomorphic processes that constantly change the landscape affect the temporal and spatial archaeological sample (Waters and Kuehn 1996). Therefore, the importance of off-site research combined with archaeological survey seems to be an important combination of strategies for fully understanding invisible and absent elements in the record.

9

The Record of Early Agriculture and its Diffusion

INTRODUCTION

The invention and spread of agriculture meant complex changes within human societies themselves and their relation with the environment. These complex changes involved processes by which resources were appropriated, thus changing energy flows within ecosystems in favor of food production. The term *Neolithic*, a name that is not used in all the world and whose usage is often questioned, broadly refers to the time frame when early forms of farming and rearing animals appeared in a particular region.

The role of geoarchaeology in the study of early agricultural (Neolithic) societies and their environments has multiple facets. First, geoarchaeologists participate in the survey and excavation of sites and the interpretation of site formation processes, site abandonment, destruction, and preservation. Second, geoarchaeology places the findings in an ecological context at the levels of setting, landscape, and environment. Third, geoarchaeology works in tandem with other fields in environmental archaeology to better understand the processes of environmental change associated with the Neolithization process, particularly assessing the impact of agricultural activities and changes on the environment, and reciprocally the environmental impact on the agricultural process itself.

As these three facets of research show, the capabilities of geoarchaeology in the field of early agricultural environments are vast. Therefore, this chapter focuses on general models for the study of early agricultural societies and their environments, highlighting the complexities and potential problems to solve in the process of geoarchaeological research, namely the points delineated in the second and third facets.

As the geoarchaeological capabilities in the study of Neolithic environmental processes are vast, so are the number of research cases from around the world. Therefore, the two significant examples selected for this chapter refer to early farming in the Near East, namely in the Southern Levant (Case 9.1), because of its importance as being one of the first cases of farming in the world and because of its continuity with the case of the already discussed Levantine Epipaleolithic (Case 8.1). To focus on the environmental aspects of non-farming societies and the process of diffusion, the second case includes the example of the arrival of pastoralism in the Kalahari and its impact on the local environment, particularly around the Okavango Delta and Lake Ngami (Case 9.2).

AGRICULTURAL BEGINNINGS: CONTEXTUAL MODELS

The biogeographic and ecological contexts

The distribution of centers of domestication has always been of interest to archaeology, anthropology, environmental history, and biogeography. The contributions of the renowned botanist, geneticist, and biogeographer Nikolai Vavilov (1887–1943) stirred the interest in understanding the geographic distribution of plant domestication, particularly in how geographic factors influenced the process of domestication in the regional context (i.e., biome, climate, and species richness). Although referring mainly to plant domestication, the ideas also permeated the biogeographic aspects of animal domesticates and their wild ancestors.

The importance of climatic conditions is often given too much importance in the processes of plant and animal domestication (e.g., Diamond 1997; 2002), sometimes overriding the influence of other factors, particularly those associated with the adaptation and malleability of genomic resources. Nonetheless, climatic influences do have interesting points of significance in the context of adaptive processes and cultural preferences. For example, it is undeniable, that the domestication of C_3 and C_4 grasses occurred in different climatic areas and that the domestication of large mammals often occurred at higher latitudes (Diamond 2002). But the cultural aspects in this relation are perhaps more important. For example, Triticeae grasses (the tribe of wheat and barley) do exist in North America, where they were never domesticated nor their wild species consumed, yet at those latitudes the tropical grass maize was preferred.

Unlike the more regionally and globally oriented biogeographic models, ecological models address more specific biotic and abiotic factors, i.e., at the level of the ecosystem. The ecological paradigm of the second half of the twentieth century influenced the idea that agriculture meant the creation of human-made ecosystems, something that is often referred to as *managed ecosystems* (Butzer 1996). Interestingly, such ecological models are more commonly adopted in archaeological and anthropological models. However, for geoarchaeology, they do have an important aspect since the ecosystem, in its cultural form, correlates better at some scales with the contextual level of landscape and at others with the level of environment.

An ecological-landscape approach implies many aspects of interest in the study of early agriculture, particularly the modalities of *landscape transformation*, an idea that at the contextual level of environment is approached as *environmental impact*, thus complementing the multiscalar view of early agricultural human-environmental relations.

Continuity and resilience are two properties that also characterize many studies of the agricultural domestication process (Rosen and Rivera-Collazo 2012). In essence, the idea of continuity and resilience suggest that the process of domestication, unlike the "success" model (*sensu* Diamond 1997), is characterized by setbacks and re-trials. Early agricultural societies, and in fact all agricultural societies, have faced setbacks created by climatic changes or by the feedbacks originated from impacts created by societies themselves on the environment (Redman 2005; Rosen 2007a; Cordova 2007; Butzer 2012).

Despite the setback in the agricultural development process, societies reorganize themselves and the system tends to continue, sometimes in the form of a resilience cycle (Butzer 2012; Rosen and Rivera-Collazo 2012). One of the better known is the first setback experienced by the earliest agricultural settlements in the Near East allegedly as a consequence of the global cooling event known as the 8.2 ka event, as represented in Case 9.1 in this chapter.

The pastoral phenomenon

Along with farming, the rearing of animals was an important phenomenon of the Neolithic process. In the Near East, plant and animal domestication appeared around the same time, perhaps implying the need to secure food and establish a sedentary way of life (Zeder 2007; Rosen 2007a). But in other cases, farming and animal rearing appeared separately, without the direct need of one for the other. In others, animal rearing appeared separately in areas where it was better fit to highly variable environmental conditions.

Animal rearing without farming often appeared in the form of nomadic pastoralism. Geographically, areas where nomadic pastoralism prevailed in environments that are semi-arid or arid (the Near East and arid parts of Africa) or areas of extreme cold (the Andes and the Eurasian far north), or a combination of cold and dry (the Eurasian steppes). Not falling into the abyss of geographic determinism, these examples are worth considering particularly in terms of identifying exceptions to the rule.

One interesting difference in terms of archaeology, and certainly geoarchaeology, is the type of relation between nomadic pastoralists and the environment, which differs considerably from sedentary farmers. Although combined forms exist, normally nomadic pastoralism tends to create a more widespread impact on the environment, which in many cases may leave an elusive mark in the record. In western Southern Africa, where pastoralism arrived in late prehistory, the impact seems elusive in some environments as they fail to appear in sedimentary records (see Case 15.2, Chapter 15). But this is not the case of certain localities where the concentration of water sources in a dry environment concentrated most of the herds (Case 9.2).

NEOLITHIC IMPACTS ON THE ENVIRONMENT AT DIFFERENT SCALES

Traditionally, most of the studies dealing with the impact of early agriculture on the environment were seen at a regional, if not local, scale. But as pointed out in Chapter 5, particularly in the context of the Anthropocene, the impact transcended the global scales, as greenhouse concentrations increased with the removal of forests for farming, the breaking of the organic soil cover, and the creation of rice paddies (Ruddiman 2014; Ruddiman et al. 2015).

Normally, at the contextual level of environment, early agricultural developments are recorded in lake deposits mainly through vegetation proxies (e.g., pollen, spores, carbon stable isotopes, and charcoal) and proxies related to water conditions (e.g., diatoms and increase of certain spores). But geomorphological records also point to changes in erosion and sedimentation. The problem is that sometimes the integration of records can show mismatches, sometimes related to dating issues, others to poor correlation between paleoenvironmental proxies, and others to the time-transgressive nature of environmental responses to cultural changes. In this kind of situation, more than a technical or specialized knowledge, an interpretive skill is necessary.

The environmental impact of early agriculture in the Near East is perceptible in pollen records in lake sediments, but the record is often blurry in time, in part because of poor dating or problems with radiocarbon dating (Butzer 2005; Cordova 2007; Rosen 2007a). Geomorphic evidence of such changes are easy perceptible in and around the sites (Cordova 2007; Henry et al. 2016), but no studies have addressed them in the broad context of the landscape (Cordova 2007). In contrast, in other parts of the Mediterranean and in Europe records of the Neolithization are much clearer in both pollen geomorphic records (e.g., Bottema and Woldring 1990; Simmons 1996; Butzer 2005; Roberts 2014). However, in these areas agriculture arrived fast enough to show its impact.

Other areas of the Old World such as East Asia have also seen studies pointing out to the modifications of geomorphic processes and hydrology of early agriculture (Rosen 2007b; 2008; Kidder 2014). However, there are challenges posed by processes in later times. Sometimes, later development blurs the effects of early pastoral/farming effects in earlier times.

However, farming and its combination with pastoralism is not the only transformer of nature. Some societies in Africa and dry parts of the Near East developed only pastoralism, which in turn has its own forms of impact, of which overgrazing is one of the most salient. Overgrazing is often conducive to erosion and sedimentation, but the selective removal of vegetation and compaction of soil, which reduces infiltration and increase of runoff. This process also implies the reduction in groundwater, which in turn reduces the level of the water table, causing streams to incise (Butzer 1982).

Evidence of early farming or pastoralism is nowhere as clear as in lake deposits, where not only changes in vegetation and soil increase sedimentation rates, but also include proxies for changes in vegetation as a result of activities (Denham, Haberle, and Lentfer 2004). Case 9.2 presents the simple case of the sudden arrival of pastoralism in the northern Kalahari, attested to in both archaeological site sediments and the sediments of Lake Ngami.

GEOARCHAEOLOGICAL CONTEXTS AND RESEARCH STRATEGIES

At the level of environment

The study of human-environmental interactions in early agricultural landscapes requires first of all an understanding of some environmental background, or more properly the biogeographic and ecological backgrounds. Thus the distinction between areas where domestication first happened (i.e., the centers of domestication) and those where domesticated plants and animals arrived by diffusion is important when assessing the human-environmental interactions and aspects such as environmental impacts.

In the centers of domestication (e.g., the Fertile Crescent and Mesoamerica) the process or processes took longer to develop, thus the landscape changes slowly and in a way become trained to domestication. In contrast, in those areas where domestication arrived by diffusion, the environment was not fully trained on specific changes, leading to a major impact, as is the case discussed above for Europe. In the latter situation, finding evidence of the earliest agricultural impact in paleoenvironmental and geomorphic records is easier.

But this is not a rule as some cases around the world show different patterns of integration of agricultural activities in the local environments. In the case of the North American Great Plains and the American Southwest native agriculture arrived and disappeared intermittently, perhaps allowing the natural systems to a regenerate. But that is not the case of European farming and pastoralism, which took place at different rates and over large areas (Beinart and Coates 2010). In such a situation, finding evidence of early farming in off-site environments is difficult, for which the focus is mainly on the record at the level of site.

At the levels of site, setting and landscape

In early agricultural contexts, sites and settings vary considerably depending on local or regional conditions such as topography, hydrology, climatic regimes, and vegetation. If one looks at the Near East, for example, sites in the early farming settlements preferred low areas and often near alluvial fans, but later settlements occupied more complex locations on slopes and hilltops, where they became architecturally more complex (Rollefson and Kafafi 2007; Goring-Morris and Belfer-Cohen 2011; Rosen 2017).

In the humid parts of the world, particularly in the tropics, preservation constitutes many problems for studying early agricultural societies. But strategies to obtain negative parts of the records are always devised. In the case of the Highlands of Papua New Guinea work has been focused mainly in wetlands where early farmers dug drainage ditches that permitted the cultivation and at the same time helped preserve the record (see Denham, Haberle, and Lentfer 2004; Haberle et al. 2012). Very important in this case is that the sediment in the ditches themselves and the wetlands provided proxies for studying the landscape before and after Neolithization.

In other parts of the world like the tropics of Central America and South America, studies have focused on the *dark earths* (discussed further in Chapter 12), which were created by early slash-and-burn farming and the determination of agricultural beginnings. In the subhumid tropics, strategies have focused on open sites and caves, as is the case in central and southern Mexico – namely the Tehuacan Valley and the Balsas Basin (Ranere et al. 2009). Alleged remains are well-preserved in caves used by early farmers as storage or trash deposits. Alluvial sediments have also provided some information for reconstructing the early farming landscape (see the case of Joyce and Mueller 1992, 1997; Borejsza, Frederick, and Lesure 2011).

It seems, however, that the characteristic of each group defines in many ways the strategies to be followed in the geoarchaeological study of early agricultural contexts, whether they refer to aspects of survey or excavation, or to the interpretation of off-site records, all of which provide material for the ecological context of early agricultural developments. The published material from different regions is vast, thus requiring perhaps more discussion in sessions dealing with regional studies.

CASE 9.1: GEOARCHAEOLOGY OF TWO NEAR-EASTERN NEOLITHIC SETTLEMENTS: AIN GHAZAL AND AIN ABU-NUKHAILA, AND THE FIRST AGRICULTURAL ENVIRONMENTAL CRISIS

The Pre-Pottery Neolithic B (PPNB) in the Levant: Environment and landscapes

The development of the PPNB (Pre-Pottery Neolithic B) in the Southern Levant (roughly 10,950–8450 cal years BP) is represented by numerous large villages, often referred to as megasites (Goring-Morris and Belfer-Cohen 2011), suggesting the advanced process of sedentism associated with the long process of plant and animal domestication (Table 9.1).

Table 9.1. Chronology of periods in the Levantine Neolithic. Based on Goring-Morris and Belfer Cohen (2011) and Banning (2012)

Stratigraphic Units	Dates cal BP	General Cultural Characteristics
	Late Epipaleolithic	
Natufian	14,900–11,750	Seasonal mobile residence, semi-sedentarism. Grass collection and gazelle hunting.
	Early Neolithic	
PPNA	12,175–11,000	Brief period in the southern Levant. First evidence of farming. Round houses. Grain storage structures.
Early PPNB	10,950–10,300	Not much different from the PPNA. Local differences. Gazelles make up for most of consumed meat.
Middle PPNB	10,150–9725	Megasites – complex houses. Goat meat replaces gazelle.
Late PPNB	9400–8900	Megasites – more complex. Goats are fully domesticated.
Final PPNB (PPNC)	9050–8450	Population centers contract. Reorganization of cultural and economic traits.
	Late Neolithic	
PNA-Yarmoukian	8400–7700	Housing less complex than PPNB. Pottery appears. Stress on farming.
PNB-Jericho IX & Wadi Rabbah	7500–6500	Increasing dependency on farming.

PPN = Pre-Pottery Neolithic; PPNA, PPNB, PPNC = Pre-Pottery Neolithic A, B, and C, respectively; PN = Pottery Neolithic; PNA, PNB = Pottery Neolithic B.

Most of these large PPNB villages occur today in areas in the Mediterranean shrubland ecozone with a few on the borders with the Irano-Turanian steppe (Figure 9.1).

Provided that ecozone limits are different today, it is obvious that the villages occupied the areas with more rain and better soils. Nonetheless, numerous sites of considerable size and complexity are found in the desert (Figure 9.1). However, the idea that most Neolithic desert sites were exclusively pastoral because they occupied a marginal area is the result of the long-believed *desert and the sown* myth, which has been debunked by recent research (Rosen 2017). The idea of marginality is difficult to assess because precipitation and vegetation were different in the early Holocene, although we are not certain where the vegetation boundaries were except for estimates provided by proxies in the sites (Cordova 2007).

The influence of a climatic shift in the decline and eventual collapse of several PPNB sites in the Levant is topic discussed for several decades (Simmons et al. 1988), and it is not until more recently that more evidence has appeared to point to the so-called 8.2 ka event as the cause of dryness that impacted agricultural activities and caused the abandonment of PPNB megasites (Robinson et al. 2005; Weninger et al. 2006).

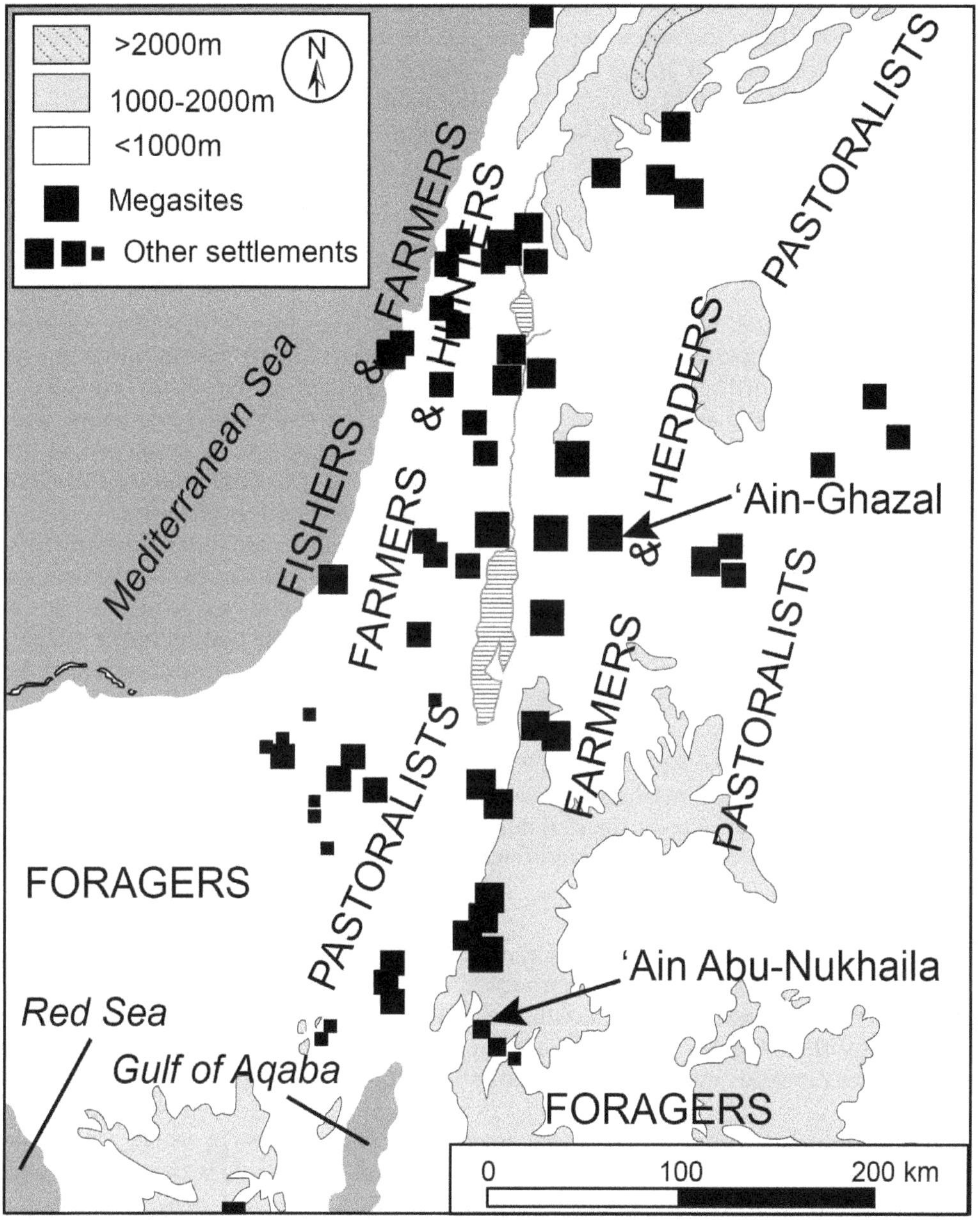

Figure 9.1. Location of most important PPNB sites in the southern Levant with the location of 'Ain Ghazal as the location of 'Ain Abu Nukhaila. Large squares are PPNB megasites. Map modified from Goring-Morris and Belfer Cohen (2011).

Surrounding this issue is the case of the 8.2 ka climatic event and the so-called "Yarmoukian (or Late Neolithic) rubble layer" (Rollefson 2009) discussed here in light of geoarchaeological research at the site of Ain Ghazal by Zielhofer and colleagues (2012). To complement the information of referring to the environmental crisis created by the 8.2 ka event on the early Neolithic communities of the region, an analysis of the geoarchaeological record at Qa' Ain Abu Nukhaila in southern Jordan is cited.

Ain Ghazal and the Yarmoukian rubble layer

The site of Ain Ghazal, located within the metropolitan area of Amman, Jordan, has been the subject of intensive archaeological and geoarchaeological research due to its importance as one of the largest villages of the PPNB (Pre-Pottery Neolithic Period B), or the time of widespread agriculture in the Levant (Rollefson and Kafafi 2007), particularly during the time when farming and pastoralism were already fully developed as the economic basis (Zeder 2011). Most of the occupation of the site corresponds to the PPNB, with traces of the PPNC and a modest occupation in the PN (Pottery Neolithic).

The site extends on a slope facing a tributary of the Zarqa River, and is built on terra rossa sediments with embedded limestone cobbles, suggesting that before the settlement was built the slopes were unstable (Simmons et al. 1988). Thus, terracing had to be implemented to support houses. Today, the stream flowing at the bottom of the valley is devoid of sediments, but it has been assumed that at the time of the occupation of Ain Ghazal a floodplain existed (Kafafi 2009; Zielhofer 2012).

Considering that 10% of the site has been destroyed, the area of Ain Ghazal is estimated between 12 and 13 ha (Rollefson et al. 1992). Ain Ghazal's occupation encompassed more than 2000 years, with its maximum extent during Middle PPNB (Rollefson et al. 1992). Abandonment and partial re-occupation during the Yarmoukian (PN) period is a topic that is of great interest in terms of the causes and processes involved, which are discussed in Case 14.1 (Chapter 14).

The broader landscape around the Neolithic settlement was a hilly area with oak woodland (Simmons et al. 1988; Rollefson et al. 1992). Many more details of the landscape around the site have been lost to urbanization, including possible evidence of the nature and extent of the former floodplain, which is assumed to have been the optimal areas for wheat cultivation (Kafafi 2009), but that view has been contested because of the lack of groundwater processes evident in the soils of some geoarchaeological profiles (Zielhofer et al. 2012) (see the trench sections studied in the maps of Figure 9.2a–b). However, these profiles are located much too high in elevation to have been reached by a high water table. Nonetheless, flood sediments are reported in trench section 3 (Figure 9.2), which suggests perhaps an extreme overbank event.

Like many other Neolithic megasites in the Levant, Ain Ghazal was also subject to shrinking after the PPNB (Table 9.1). Occupations after this time was modest (PN-Yarmoukian), and the site practically never recovered to its former maximum size in the PPNB. One problem that has been pointed out at this site, as well as other megasites, is the presence of a rubble layer, which has been controversial (Rollefson 2009). The consensus is that the rubble layer has been caused by processes such as landslides and colluvium caused by the abandonment and degradation caused by human activities, presumably grazing, on slopes.

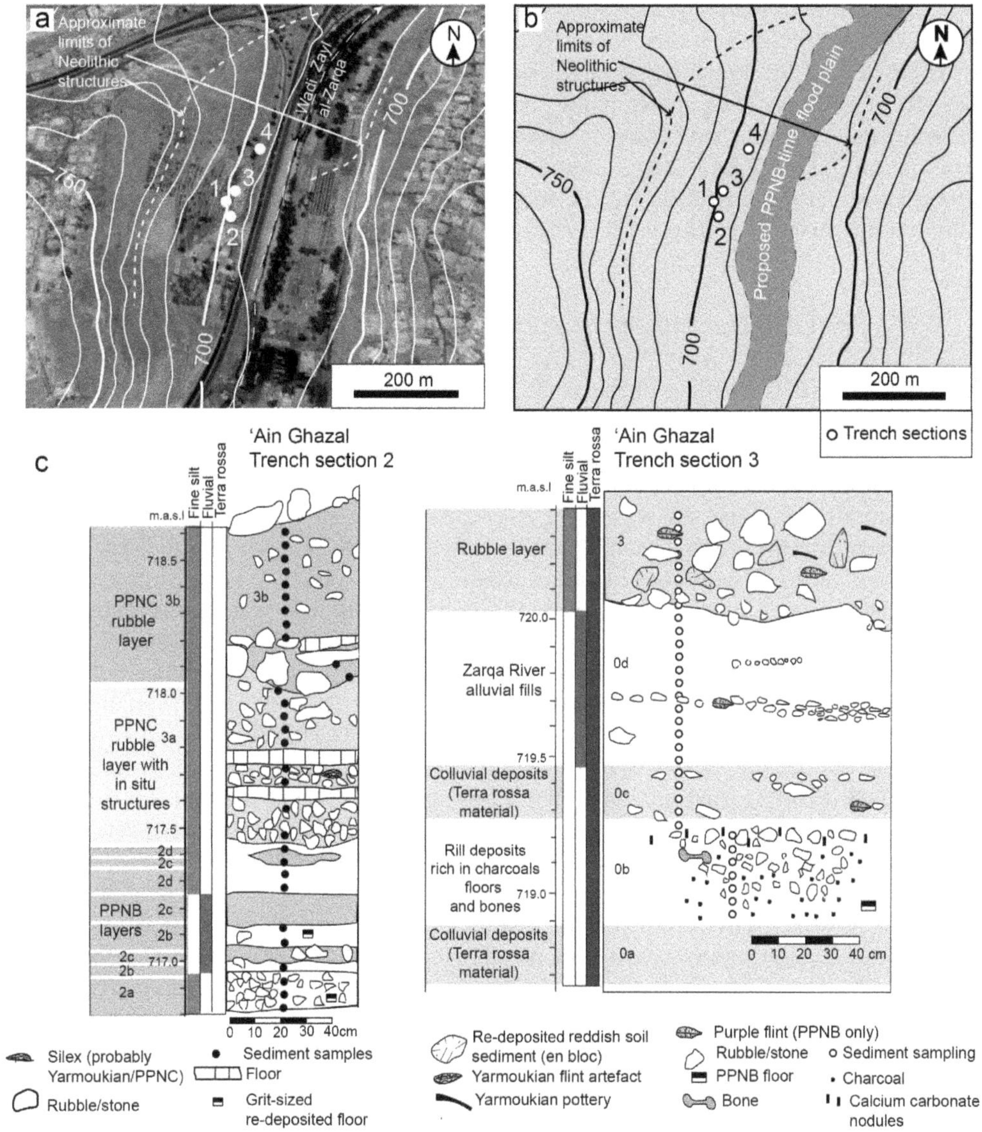

Figure 9.2. a-b) Ain Ghazal Area with approximate location of the trench section referred to in text. Map based on Rollefson (2009) and Zielhofer et al. (2013); c) summarizes stratigraphy of two selected trench sections (sections 2 and 3 on maps above). Modified from Zielhofer et al. (2012).

At Ain Ghazal, this hypothesis has been challenged by a geoarchaeological study that combined several techniques to test the different possible origin of sediments in the matrix that comprised the putative rubble layer (Zielhofer et al. 2007). The rubble layer was highly variable as it can be seen in some of the profiles (Figure 9.2). The study concluded that the nature of the layer varied, and there is no evidence to link it to anthropogenic causes, i.e., post-abandonment land degradation, nor to the abandonment itself. In fact, the number of fine particles derived from dust suggest that the processes were climatic.

The results of the study, whose methodology is unquestionable, represents only a small portion of the site. Thus, generalizing the results for the entire site and all the PPNB is difficult, unless similar studies are implemented at all other PPNB megasites. In any case, whether caused by climatic deterioration, or the lack of maintenance of terraces, or as also proposed, anthropogenic degradation, the rubble layer is an indication of an environmental crisis, the first one in the agricultural history of the world.

Ain Abu-Nukhaila and the invasion of sand on a small wetland

Located in one of the valleys of the Wadi Rum, Ain Abu-Nukhaila is a small settlement of about 1200 m^2 with occupation over a brief period of about 200 years in the Middle PPNB (Henry and Beaver 2014; Henry et al. 2016). The site is located on a ramp formed by a former alluvial deposit with eolian component (Cordova, DeWitt, and Winsborough 2010). After abandonment, the site was never re-occupied. The spring, where the site takes its name from, however, attracted Nabateans, travelers and nomadic herders of different periods, but no post-Neolithic settlement exists.

West of the site, a small playa known as Qa' Ain Abu-Nukhaila was studied as part of the geoarchaeological research of Ain Abu Nukhaila's environmental context (Figure 9.3). The stratigraphy of the Qa' profile (Figure 9.4) shows cycles of predominantly eolian sand deposition (dry periods), alluvial periods (wet-dry with torrential rains), and laminated silt (wet and more stable). In the entire sequence, corresponding to 11 ka to the present, the wettest and most stable period is characterized by the deposition of the sediments of layer 9, dated around cal 8755 BP with an error age of 550 (Cordova, DeWitt, and Winsborough et al. 2016). The layer also shows evidence of a more vegetated desert and strong evidence of pastoral activities and even the presence of domesticated wheat, biomass burning (Henry et al. 2016).

The question as to why domesticated wheat in a site previously believed to be pastoral because of its location in the desert is not clear, but the only possible reason is short-lived, but a period of enhanced precipitation. One possible reason for the local wet conditions may be the enhancement of water from the spring fed by rains in the Jabal Ramm Mountain (1112 m), which may have increased the flow of water and the possibility of a more permanent wet area (Cordvova, DeWitt, and Winsborough 2014). This may be an opportunistic circumstance that could epitomize the productive function of other PPNB sites in the desert.

The study of sediments in Qa' Abu Nukhaila provides not only an off-site record of the local environment through the modes of sedimentation, but also a source of information (e.g., pollen, phytoliths, spherulites, and other proxies) for reconstructing ecological aspects of the environment (Henry et al. 2017). In turn, the sequence of layers 7 to 9 show parallels to the short existence of the Middle PPNB occupation of the site nearby, but also provide evidence of an abrupt climatic change, likely 8.2 ka, which may explain not only the lack of re-occupation of the site, but also a record to the widespread climatic crisis that affected PPNB sites in the broader southern Levant.

Data from Qa' Ain Abu-Nukhaila seems to confirm many facts about the end of the Early Neolithic in the Levant and the 8.2 ka event. One is that it shows the abrupt change marked by the drying of a wetland (layer 9) and the deposition of shifting sands (zone 9),

Figure 9.3. a) View the landscape context of the PPNB site of Ain Abu Nukhaila, a view from the stratigraphic site at Qa' Ain Abu Nukhaila; b) with stratigraphic profile showing the top 100 cm of the stratigraphic section (Figure 9.4) during the collection of samples at 5-cm intervals.

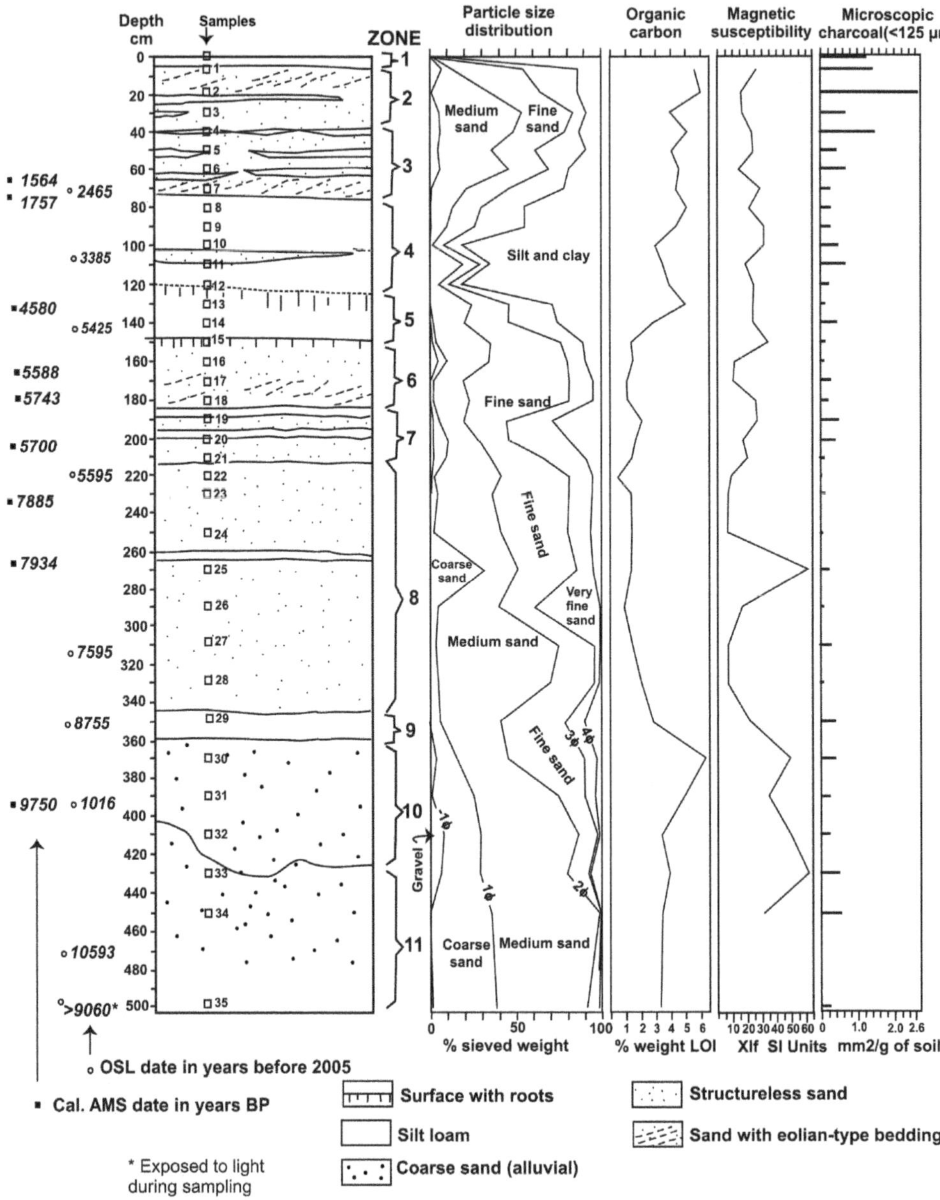

Figure 9.4. Qa' Ain Abu Nukhaila stratigraphy. Modified from Cordova, DeWitt, and Winsborough (2014).

which lasts for at least two millennia. This supports what Zielhofer and colleagues (2012) provide about dust deposition after abandonment. However, Qa' Abu Nukhaila also shows evidence of human environmental impact, despite being a small and short period, which also confirms the magnitude of the initial impact of plant and animal domestication, at least locally.

CASE 9.2: THE ARRIVAL OF PASTORALISM AROUND LAKE NGAMI: RECORDS FROM SITES AND LAKE SEDIMENTS

The appearance of pastoral societies in arid Southern Africa occurred ca. 2000 years BP, a relatively late date compared to other parts of the continent. Hypotheses to explain this delay in diffusion range from persistent aridity earlier in the Holocene to the problems the tse-tse fly posed for pastoralism to spread across tropical Africa (Gifford-Gonzalez 2000). Nonetheless, once the diffusion of pastoralism reached the southern limits of the tse-tse fly zone, approximately in the location of the Okavango Delta, pastoralism spread south quickly into the rest of present-day Botswana, Namibia, and western South Africa. The overlap of radiocarbon date standard deviations between the Lake Ngami region and the Cape region of South Africa suggest that the dissemination of pastoralism took less than 200 years to reach South Africa's south coast (Sadr 1998; Robbins et al. 2008).

The Okavango Delta occupies an area of the northern Kalahari where faults have created a series of lacustrine basins and river paths that deposit alluvium over a vast area (Figure 9.5). The region is occupied by various types of dry savanna on former longitudinal dunes and alluvial and lacustrine plains. Although apparently dry (precipitation ranging between 400 and 450 mm a year in the summer), the grasses associated with *sandveld* savanna and the humid areas of the delta and alluvial plains provide a great amount of food for livestock. In particular, the area around Lake Ngami (Figure 9.5) provides several grazing spots in alluvial and lacustrine plains, small ponds and playas, the sandveld, and the diverse lithological grounds of the Ghanzi ridge to the south and the hills and pans to the northwest (e.g., Tsodilo Hills and adjacent lacustrine pans).

The earliest pastoral sites discovered in the area are located around the modern settlement of Toteng, located at the confluence of the Nchabe and Kunyere rivers (Figure 9.5). For the most part, the Kunyere feeds Lake Ngami with water from the Delta, while the Nchabe drains it into the Boteti River, which in turn drains water into the Magkadikadi Pans (Figure 9.5). At times, however, the Nchabe can feed water into the lake if there is an excess of water from the east via the Taklamakane River. In the past, other feeder channels emptied water into Lake Ngami from the north, a process that ceased in recent historic times (Shaw 1985; Shaw et al. 2003).

In this region, archaeologists have also studied the transition of sites from foraging to pastoralism. One interesting aspect is Toteng 7, a site with strong evidence of foraging and fishing from times shortly before the arrival of pastoralism (Figure 9.5). The site lies in a terrace that is at one of the highest stands of the lake 4000–3500 BP (Robbins et al. 2008). This period coincides with evidence of an improvement in vegetation, particularly towards trees and grasses (Cordova et al. 2017).

Sites with remains of domesticated animals, and later pottery, also coincide with another wet phase, with a lake stand that was high but lower than the previous level. The terrace

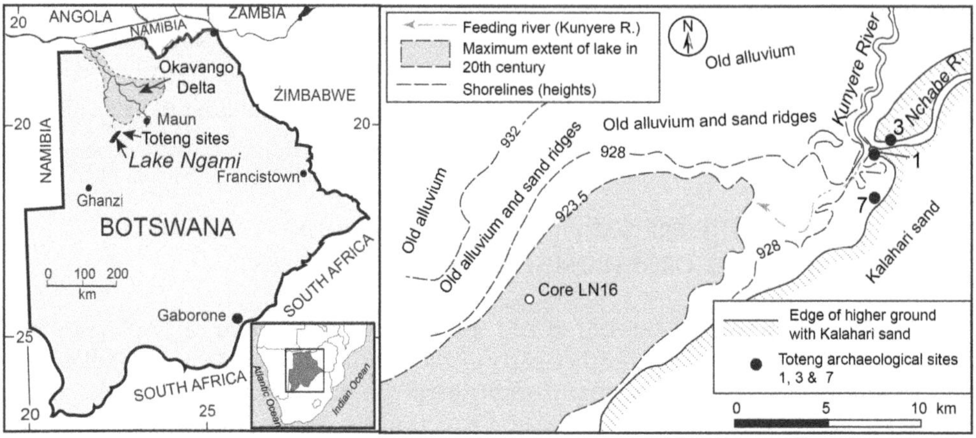

Figure 9.5. Lake Ngami with paleo-shores, location of core LN16, and Toteng sites 1, 3, and 7. Modified from Shaw, Bateman, and Davies 2003, with information from Robbins et al. (2008).

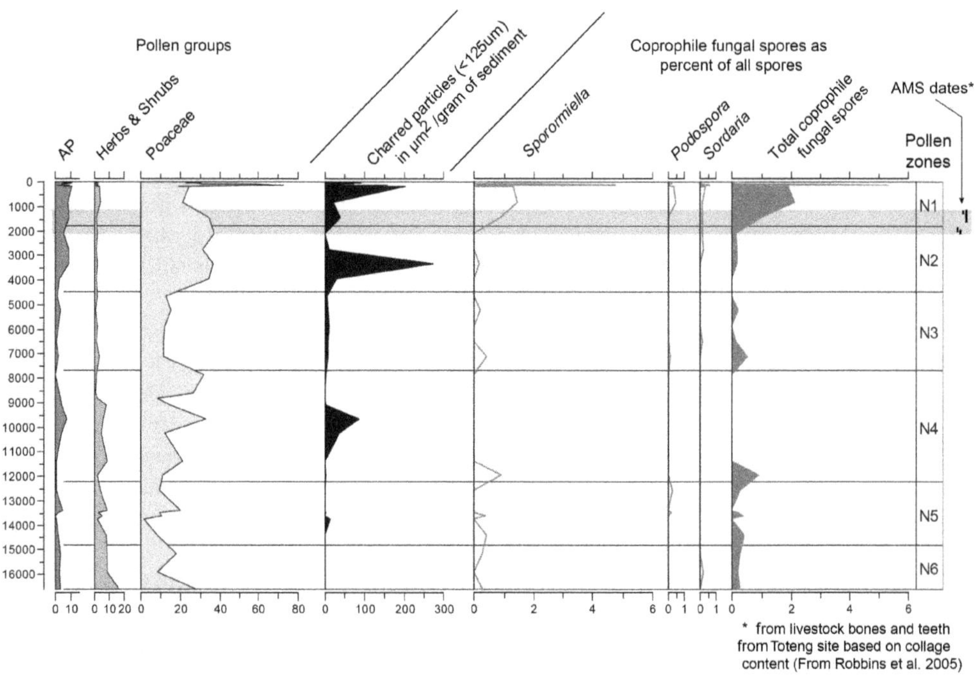

Figure 9.6. Pollen summary, coprophile fungal spores, and burning proxies obtained from core LN-16 in Lake Ngami. Based on Cordova et al. (2017).

was occupied at this time by cattle and sheep herders (sites Toteng 1 and 3), which also suggests a relatively wet period, this time with an estimated lake level stand at 934 m. Recent pollen research suggests a wet period around 2000 BP (Cordova et al. 2017). The important aspect to recognize is that there are no structures in the sites, suggesting that these were seasonal encampments, perhaps some form of nomadic pastoralism.

The impact of early pastoralism on the soils and hydrology around Lake Ngami is not known. Geomorphic evidence is difficult to assess because this is a flat area with predominantly sandy soils. However, evidence of erosion may be interpreted from the lake deposits and its proxies. Sedimentation rates increased from 0.1 mm/year to 0.9 mm/year after cal. 2000 BP, and then to 3–4 mm year after cal. 212 BP (Cordova et al. 2017).

A spike in magnetic susceptibility (indicative of high erosion of organic matter from soils) also occurred around 2000 BP (Figure 9.6). This increase is accompanied by large concentrations of of coprophile spores (mainly *Sporormiella*, *Sordaria*, and *Podospora*), and evidence of burning of the general vegetation (charred particles) and grasses (burnt-grass phytoliths) (Figure 9.6). Finally, vegetation changes during the "pastoral phase" are difficult to discern in the pollen record, but the Acacia (*Vachellia*) pollen suggests disturbance (see Cordova et al. 2017). Among other evidence of the pastoral period is the notable decrease in diatoms, suggesting possible changes in the limnnologic conditions of the lake.

The paleofire proxies, however, are difficult to use as evidence of the arrival pastoralism. In fact, the largest peak of burning (charred particles) occurs actually in the period 4000–3000 BP, predating the pastoral period. This phase, however, coincides with regional levels of moisture, and high lake level stands, suggesting that burning occurred as vegetation became denser and fuel for fire increased. Thus, the true influence of pastoralism, in addition to fire proxies, should include all other proxies (pollen, spores, diatoms, sedimentation rates, magnetic susceptibility, etc.).

Periods of substantial drying of the lake exist during the past 2000 years, particularly in the past few centuries (Shaw et al. 2003; Burrough et al. 2007), including one that occurred after the visit to the area by missionary and explorer Dr David Livingstone. During these dry periods, the wet lake bed covered with a grass savanna was attractive to cattle, thus explaining perhaps the strong compaction in the sediments in the top 80 cm of the core.

In summary, the time of arrival of pastoralism recorded in the archaeological sites seems to match with changes in various proxies from the nearby Lake Ngami. However, questions about the incidence of fire, as well as putative erosional processes in the area, are still not clear. Despite the clear correlation of records in this area, not all lacustrine records in the broader Southern African region attest to the arrival of pastoralism (further discussed in Case 15.2). Therefore, perhaps finding the right proxies and a more focused geoarchaeological research protocol could target impacts that are hidden in the landscape.

10

Complex Societal-Environmental Systems and the Collapse Phenomenon

INTRODUCTION

Inside and outside the archaeology circles, the views on society-environment interactions have been multiple and sometimes supported by opposing points, particularly with regard to understanding societal success, stability, demise, and resilience in relation to environmental change. In this rational context, geoarchaeology, with its methodological and technical capabilities, becomes a key component in the effort to search and find evidence of the environmental factors that interact with societal development.

The literature on societal-environmental interactions is so rich in examples of geoarchaeological research that it would be difficult to discuss many of them within the limits of a chapter. Thus, this chapter sets only the basic framework for complex societal-environmental systems, focusing on the most important aspects of research in the topic in the twenty-first century. Such topics include the rise and collapse of complex societies and aspects of sustainability, vulnerability, and risk. Other aspects related to environmental factors in the development of complex societies are discussed in Chapters 11 (ancient rural landscapes), 13 (natural catastrophes), 14 (environmental crises), and 17 (the geoarchaeology of the contemporary past).

In the present chapter two cases of geoarchaeological interpretation of complex environmental and societal systems are included. One is the environmental context of three Old-World, early hydraulic civilizations from the viewpoint of sustainability and resilience (Case 10.1). The second is the role of geoarchaeological research in the highly popular case of the collapse of the Classic Maya (Case 10.2).

COMPLEX SOCIETAL-ENVIRONMENTAL SYSTEMS

In archaeology, the idea of *complex social systems* been the pivotal term to explain cultural or social development through time, particularly referring to the transition from mobile societies (i.e., foragers) to sedentary and then to complex hierarchical systems (see examples in Bentley and Maschner 2003). This idea has its foundation in the concept of *complex society*, which refers to social, political, and economic process that sustain hierarchies and institutions vital for the formation of states. However, this linear system has been criticized because of its linearity and simplicity, or because it does not necessarily reflect the societal relations, as is the case of complexity in hunter-gatherer cultural development (Holdaway, Fanning, and Rhodes 2008; Smith 2012; Johnson 2017).

In geoarchaeology, the idea of complexity encompasses several systems, including the geomorphic systems (e.g., Table 3.2); the environmental systems determined in part by geographic and ecological systems; and by the same general social complexity system defined by archaeologists. These interactions are in their own sense complex because they bring in the complexity from both the societal and environmental sides, an idea that has recently been influenced by the ecological and societal models (Scheffer 2009). Therefore, the interaction has to be seen as a *complex societal-environmental system.*

The number of topics concerning complex societal-environmental systems (henceforth CSES) is vast and the number of examples in the literature has increased in recent years. But one that has received more attention in archaeology in the past 10–15 years is that of *societal collapse* (Schwartz 2006; McAnnany and Yoffee 2010; Butzer and Endfield 2012; Johnson 2017; Middleton 2017), a topic that has involved geoarchaeology along with a variety of other fields in the search for the causes and patterns of change that lead to the demise of past societies. Linked sometimes to the processes of societal collapse are concepts such as vulnerability and risk, and sustainability, which are discussed briefly below. Associated with societal collapse is the case of environmental crises, which is a topic treated separately in this book (Chapter 14).

THE COLLAPSE PHENOMENON

The explanation of rise and collapse

The interest in civilization collapse is as old as archaeology, particularly because the idea that people who built monumental buildings one day ceased to do so. Today, the topic has received so much attention by archaeologists that it is becoming almost a field within archaeology informally known as "collapsology" (Middleton 2017). However, the topic also attracts the interest of historians, geographers, and other scholars with interest in environmental change. But with so many cases of past civilization collapse around the world, the different disciplines involved, and their different schools of thought, the interpretive models to explain collapse are multiple.

In the course of the twentieth century, the processes of rise and fall of civilizations have been seen as a continuous cycle, as was once proposed by Russian ethnologist Lev Gumilyov (Gumilev) (1912–1992), whose theory, called *Passionarity* (*Passionarnost'* in Russian) *and Ethnogenesis*, aimed at explaining rise and fall of an *ethnos* (i.e., civilization) (Gumilev 1990). In the Soviet academic circles his theory was coldly received by historians and archaeologists, but embraced by geographers. In this sense, geographers seem to be more comfortable with the idea of cycles, fluctuations, or oscillations, in part because many aspects of their study (e.g., climate) occur cyclically.

Among other models, Colin Renfrew's (1984) considers several forms by which civilization collapse can occur: collapse of central administrative organization, collapse of the traditional elite class, collapse of the centralized economy, or just settlement shift and population decline. Although this model has provided the socio-political bases for more recent models, it, however, places the causes of collapse within societies themselves, often bypassing the possible environmental factors.

Joseph Tainter (1988) assessed many of the ideas and models that previously had touched on the subject of complex society collapse, pointing to the importance of environment, but also proposing that natural catastrophic events, to which societies can be powerless, often play an important role in their collapse. However, the idea of natural catastrophes as drivers of collapse can be sometimes tricky and variable according each case. Thus, catastrophic events are an aspect that undermines the concept of failure to adapt, so persistent in many cultural models, and in particular in Jared Diamond's (2003) *Collapse* book. Thus, the term collapse, and its catastrophic meaning, created controversy, particularly when in some cases societies have gone through regeneration processes (Cooper 2006).

The idea of societal collapse has stirred controversy, in part because many authors are not clear on what they mean by "collapse" (Tainter 2008), in part because of the multiple variables to be taken into account when explaining it (McAnnany and Yoffee 2008; Middleton 2017), and also because many cases of collapse from around the world are judged by the same standards (Tainter 2008; Butzer 2012).

Generally, *societal collapse* means reduction of population and simplification of social complexity, or degradation of social cohesion, values, and well-being (Diamond 2003; Tainter 2008), and often political decentralization and fragmentation of a once-powerful state (Butzer 2011). The term sometimes is replaced by *fall* (as in the rise and fall of civilization) or *decline*. In some cases, natural catastrophes are involved, precipitating a rapid collapse, as opposed of environmental crises, which are more complex and protracted processes that also lead to decline.

Another controversial aspect of the study of collapse is its causality, which often swings between internal (societal, political, or cultural) factors, and external ones, or those out of society's control (e.g., droughts, earthquakes, volcanic eruptions, flooding, etc.). The emphasis on culture and or administrative failures has been of importance in many models, particularly those that follow Renfrew's (1984) collapse model, but at the other end of the spectrum, certain models seem to be more inclined to environmental aspects, going to the extreme to blame a particular non-cultural phenomenon (e.g., climate). But there are also those that combine both, particularly in cases where environmental crises are identified in the records.

One has to acknowledge also that models of societal collapse are subject not only to the interpretation of the archaeological and environmental record, but also to the availability and resolution of such records. Thus, cases where environmental data is invisible, missing, or, worse, not properly sampled or recorded, the interpretation leans towards the internal causes. This was for many decades the case of the Classic Maya collapse, whose original causes were attributed to wars, until environmental data proved that collapse was far more complex than just warfare.

Despite the many different cases of societal collapse in the archeological and historical record, there is a trend to find a common denominator in the causes and effects of collapse. This seems to be an error that leads only to more contentious discussions, which only divert from the idea that each case has to be analyzed in its own context. Thus, instead of searching for universal models to explain collapse, perhaps a model that stresses on divergence from general models should be created.

Although the topic of societal collapse has been discussed in earnest since the decade of the 1980s, the decade of the 1990s saw an increase in studies on this topic in part because of the paradigms appearing around global environmental change, the disintegration of states in Eastern Europe and the former Soviet Union, and the now more convincing climatic models (see Chapter 5 for this paradigmatic phenomenon).

The first decade of the century saw a rapid increase in papers associated with the topic of collapse. In part, this occurred as a response to Jared Diamond's (2003) *Collapse*, in a series of meetings and sessions where scholars from different disciplines discussed the topic (Costanza, Graumlich, and Steffen 2007; Railey and Raycraft 2008; McAnnany and Yoffee 2010; Butzer and Endfield 2012; among others). The meetings were significant in the sense that the entire idea of complex societal collapse was discussed, and new proposed ideas for collaboration, interpretation, and diffusion to the public were brought to the table.

Environment and societal stability

One problem that persists in the interpretations of the archaeological record is that there is too much attention on the rise and collapse of civilization, but little on how complex societies achieve a stable state that keep them from collapsing. The term *success* is often mentioned in the literature, particular in the popular literature on past civilizations instead of the more complex, but more accurate term *stability*. But behind success there is a stable system, which can be explained as a resilient socionatural system (van der Leeuw 2009).

In his 2012 essay on the environmental and societal factors of collapse, Karl Butzer underlined the nature of societal collapse using five different examples, which led to the conclusion that of all the factors involved (institutional failure, civil war, climatic forcing, demographic retraction, and ideological change), institutional failure was the one that characterized most cases (Butzer 2012). In the same essay he stresses the concept of social stability as the key to maintaining a state from collapse, a process in which resilience is one of the key factors.

There is a tripartite system of resilience that includes environmental, political, and social forms of resilience, which according to K.W. Butzer (2012) negates the idea of collapse, and instead brings up the idea of evolution and change. This tripartite alliance is rather seen in terms of stability, which in turn is broken into environmental, political, and social stability. However, these stable systems are not permanent, but constantly changing and adapting to new external (environmental) and internal (societal) changes. Failure to adapt to such changes can lead to a vulnerable state, which can put one, two or all of the tripartite alliance elements of stability at risk. Thus, if a catastrophe or abrupt climatic change strike, the system may go into a crisis. The case of the Classic Maya was one example of a tripartite system that became so vulnerable by the time precipitation became highly variable (Figure 4.2).

Sustainable systems, risk, and vulnerability

Among the many aspects that are often brought to the discussion of collapse, are the concepts of sustainability, risk, vulnerability, and other terms related to coupled human environmental systems (see Kasperson, Kasperson, and Turner II 2010). Although these terms are broadly understood in environmental science, ecology, environmental archaeology, and most social sciences, the problem comes when trying to interpret them in the archaeological record. Even in the geoarchaeological record, where both human and environmental components are examined, it is difficult to discern them, particularly when the record is fragmentary (Doolittle 2006).

Sustainable systems in modern traditional societies are common, but they can lead to biases when interpreting past systems. Thus, what may look sustainable today may not have been sustainable in the past (Fisher 2009). In other words, "sustainable systems" might have then a fleeting meaning, which can only be securely labelled "sustainable" against its social and environmental context. In fact, the reconstructions of the use of the three rivers (Case 10.1), which today are managed very differently with modern technology, had to draw on modern analogs or recent historical examples.

There is also the issue of sustainability and scale, meaning that some systems may look sustainable at a particular scale, but not at others. Thus, cases such as raised fields in the tropical lowlands may be sustainable in their own context, but perhaps not in the context of modification of hydrology, vegetation, wildlife, and nutrient cycles. Such subtleties sometimes, then, render the term "sustainable" a relative one (Doolittle 2006).

In some scholarly circles, particularly in the past century, land degradation has been associated with the Malthusian ideas of population growth popularized by population models developed by economist Esther Boserup (1965). Such ideas were characterized by the linear relation between population and environment. However, numerous archaeological examples of complex societies show that population density does not always have positive feedback relations with land degradation (Brown 2017). It is usually technology, settlement pattern (intensive or extensive), and bureaucratic administration, and overall functionality of systems, that under conditions of high population density help prevent the environment from degrading (Scarborough 2009; Butzer 2012; Kennett and Beach 2013).

Population growth and land degradation do have a positive relationship in cases such as the early colonization of previously unoccupied lands (e.g., Iceland and Rapa Nui) or early agricultural expansion, often during the Neolithic period, as is the case of northern and Eastern China (Zhuang and Kidder 2014; Rosen et al. 2015).

Thus, the idea that large population densities practice exploitation of resources in an unsustainable way is also a myth that is challenged by case studies from around the world (Scarborough 2003). One example, commented on in Case 10.1, suggests that even under high population density the Nile flood plain was managed sustainably (Butzer 1976). However, these details are sometimes hidden in the paleoecological and geoarchaeological records, which in turn requires a look at aspects beyond urban settings.

Geoarchaeology and the "collapse" question

As geoarchaeological research is a key player in the fields that reconstruct paleolandscapes and paleoenvironments surrounding past societies, it certainly plays a key role in studying cases of societal collapse. One interesting aspect that is often brought up in geoarchaeological research is the concept of scale, which in many ways is a key to solving the complexities of societal collapse. The importance of scale in the search for the causes of societal collapse can be seen in this statement: "Indeed, studying collapse is like viewing a low resolution digital photograph: it's fine when small, compact, and viewed at a distance but dissolves into disconnected parts when examined up close" (McAnnany and Yoffee 2010: 5).

In the early days of archaeology, societal collapse was studied mainly at the site level, particularly in urban sites, with little effort to look at the broad picture. For example, the Maya collapse was seen more as a function of the abandonment of temples and palaces, and not as a function of the economic means that sustained the complex societies that built and sustained them. Yet, the interpretations did not make sense, as the real causes of the abandonment were not satisfactory. It is not until the focus began to be directed to the countryside, away from any monumental architecture, to look at soils, sediments, and the morphology of the landscape with its *landesque infrastructure*, that the picture became clearer (Case 10.2).

As research began to focus on the countryside, it also began to look at phenomena acting at broader scales (e.g., effects of climate shifts, intensification of weather phenomena, and volcanic eruptions), thus providing an environmental background to the local and regional events. The collapse of the earliest villages in the Near East (Case 9.1), and the collapse of the Early Bronze Age societies in the Near East at the end of the third millennium BC (Case 14.1), among others, show the importance phenomena at large-regional and global scale can have on developments leading to climatic deterioration linked to collapse.

RESEARCH CONTEXTS

Geoarchaeology of urban contexts

The urban environment, understood in the sense of agglomerated settlements, has its own archaeological complexity, particularly when it comes to defining sequences of events (Harris 1989) and formation processes (Schiffer 1987; Goldberg 2006). However, complexity is also dependent on the type of urban context where research takes place, of which three situations can be distinguished.

One, he urban context of an abandoned site, which includes mainly the study of ancient cities with no associated modern settlement. One situation takes place when an ancient urban settlement is abandoned and never reoccupied by a later or modern urban settlement, as is the example of Teotihuacan in Mexico. The second situation occurs when an urban settlement has been continuously occupied, thus having multiple occupation layers to the present. This situation prevails in most long-standing cities, some of which have a great deal of research, as occurs in many European cities. The third situation exists when urban context in which the focus of the research is not necessarily urban layers, but cultural, or even natural, features that predate urban construction.

The distinction of the three urban geoarchaeological contexts above is meant to distinguish the circumstances of research and the methods and approaches used. In situation one, projects usually involve many seasons, allowing planning, long-term hypothesis building and solving, and the time required to focus on details.

The cases of situations two and three are mostly carried out through salvage or CRM archaeology because usually findings appear as construction or city projects develop. There are a few cases, in which the importance of the project or the circumstances of research allow for longer-term projects, as was the case of the Globe Theatre in London, or the more permanent Great Temple project in downtown Mexico City.

Sometimes there are combinations, say of cases two and three, as in the study of cores along 2nd street in Manhattan, where J. Schuldenrein and M. Aiuvalasit (2011) investigated objects of historic New York, the native or pre-European local settlements, and even features created by the glacial and deglacial phases of the lower Hudson Valley during the terminal Pleistocene. The study also revealed aspects previous to the arrival of the Europeans and the expansion of the city from its original location in lower Manhattan to the north.

Interestingly, most advances in urban archaeology have been carried out under circumstances of rebuilding after destruction, either by earthquakes or war. This has been the case in many English cities, whose urban centers had to be rebuilt after the bombings during WWII. A more recent example is the work of excavating and mapping the Roman port of Berytus, as part of the reconstruction of the center of Beirut after the Lebanese civil war, a study that has involved geoarchaeological research (Marriner, Morhange, and Saghieh-Beydoun, 2008).

Regardless of the circumstances of the finding, the important methodological aspect of urban centers is that the stratigraphy and layout show tremendous complexity. There are often intrusions, missing layers, contamination, pedological features, and other features sometimes interbedded with destruction and deposits of natural systems (e.g., flood deposits, volcanic ashes, or dust). Thus, this environment, in addition to special techniques requires special interpretive approaches (Rosen 1986; Courty, Goldberg, and MacPhail 1989; Goldberg and MacPhail 2006; Butzer, Butzer, and Love 2013).

The non-urban contexts

The non-urban contexts that form the complex societal-environmental systems (CSES) include rural contexts (the topic of Chapter 11) and other contexts that may or may not be agricultural (necessary for rural) but overall fall within the broader area of *hinterland*. Other non-urban contexts include shipwrecks, port facilities (when they are not part of urban contexts), dikes that are not meant to be of agricultural use (e.g., the flood protection dikes in Lake Texcoco, Mexico), defensive structures (e.g., the Wall of China and Hadrian's Wall), roads and trails (e.g., Roman roads and the Maya *sacbeob*), and recreational areas for the elite (e.g., gardens, baths, and villas). Despite the diversity of non-urban contexts, perhaps the most notable to single out is the rural context, in part because of its significance in sustaining society in general and in terms of concepts related to the CSES, e.g., sustainability, risks, and vulnerability.

Obviously, it is not always easy to separate the urban from the non-urban contexts. At the site, setting, and often at the landscape level it is easy to separate them, but not always at the environmental level. At the landscape level it is perhaps useful to try to delineate the urban from the rural, depending on the case. Thus, the idea of separating other aspects such as rural landscapes (Chapter 11) is more technical than integrative.

CASE 10.1: ANCIENT SUSTAINABILITY, RISKS, MANAGEMENT, AND CENTRALIZATION IN LARGE RIVER BASINS: THREE EXAMPLES

The so-called "hydraulic societies" was a term conceptualized by the anthropologist Karl Wittfogel (1896–1988) in his Oriental Despotism theory to refer to Old-World societies that depended on large-scale irrigation systems, such as those in Egypt, Mesopotamia, and the Indus Valley. Wittfogel's ideas were heavily criticized by anthropologists, archaeologists, and various social theoreticians, leading to the dismissal of the term hydraulic society, which is hardly ever used at present.

Although the term *hydraulic society* has fallen into disuse, many aspects of ancient hydraulic civilization have been taken for granted despite their importance in aspects of environmental crises and collapse (Butzer 1997; Hassan 1997; Butzer 2012; Kidder and Liu 2014). Interestingly, some issues originally studied in the context of those societies, which coincided with the time geoarchaeology was becoming a structured field, are still lingering, particularly in the context of current human ecological aspects (Butzer 1982; 2008).

Thus, instead of referring to the Wittfogelian model of hydraulic society, here the concept of large-river SCES is briefly reviewed in the context of the modern discourse of societal collapse, focusing on aspects of sustainability, vulnerability, and risk. To do so, three large-river cases are reviewed: the radial irrigation system of Mesopotamian rivers, the basin systems of the lower Nile Valley, and the complex canal system of the the lower Yellow River Valley.

The Mesopotamian *radial system* of irrigation was meant to direct water from the main channel of the larger rivers into the fields, diverting it into smaller and smaller canals (Figure 10.1). The diversion generally occurred from a stream channel higher than the adjacent irrigated surfaces, thus utilizing gravity. This could have happened only when the river was at high flood, normally after the winter rains. The system was highly dependent on rain and snow melting in the mountainous upper part of its catchment in present Turkey and Iran. The fields were in areas with a high water table, which represented a risk for salinization.

The Egyptian *basin system* of irrigation (Figure 10.2) was meant to direct water from the Nile at high flood into fields in basins. It was highly dependent on monsoon rains in the upper catchment in tropical Africa, which by the end of the fall had already reached lower Egypt.

Although K.W. Butzer (1976) seems to agree that the managing of the Nile was sustainable, the situation was in part more circumstantial. The sustainability of the Nile was also related to the tighter control of the system by the Egyptian Pharaoh, as opposed to the looser control of the more fragmented Mesopotamian administration, sometimes divided into different states along the basin.

The lower Yellow River Basin in northern China was one of the centers of ancient civilization, with evidence of cereal cultivation beginning in the middle Holocene. One important environmental aspect in the functioning of this region's agriculture is the dynamics of the Asian monsoon, whose retraction and intensification mean drought and flooding, respectively. To minimize the risk of drought and flood, dams, canals, and other works were built. Additionally, to protect fields and cities from flooding, levees, dams, and canals were also constructed.

Despite all the structures built to minimize the effects of floods, heavy loads of sediments became more and more difficult to control. Heavy loads are particularly common in this region, because the Yellow River's system drains the soft and easily erodible edges of the Loess Plateau (Figure 10.3). In ancient times, large volumes of sediment, carried particularly during flood events, caused overbank accumulation on cultivated land and towns. Additionally, they caused channel *avulsion* and the progradation of the delta into the Bo Hai Gulf (Saito 2000) (Figure 10.3b).

Channel avulsion encompasses the migration of channels across the flood plain resulting in a readjustment of the hydrological network. Extreme events in the lower

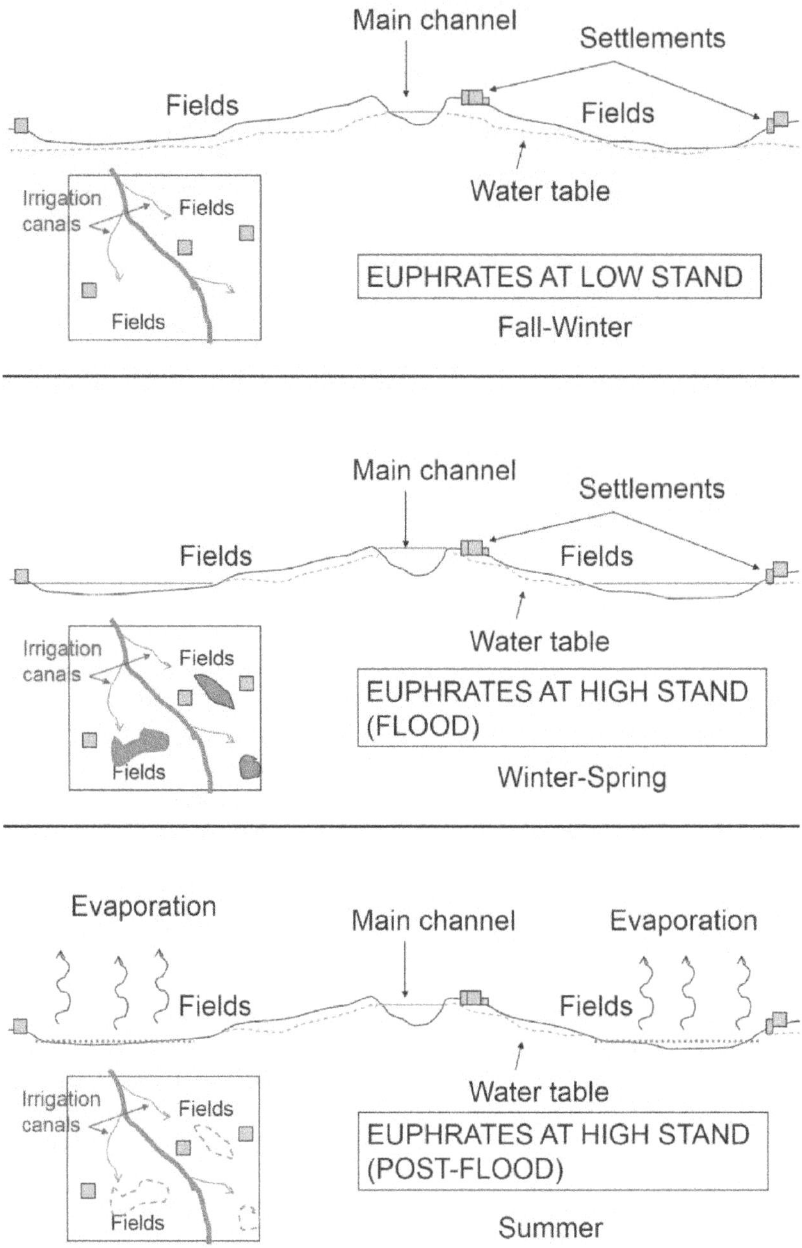

Figure 10.1. Mesopotamian radial irrigation system. Based on description in Butzer (1976).

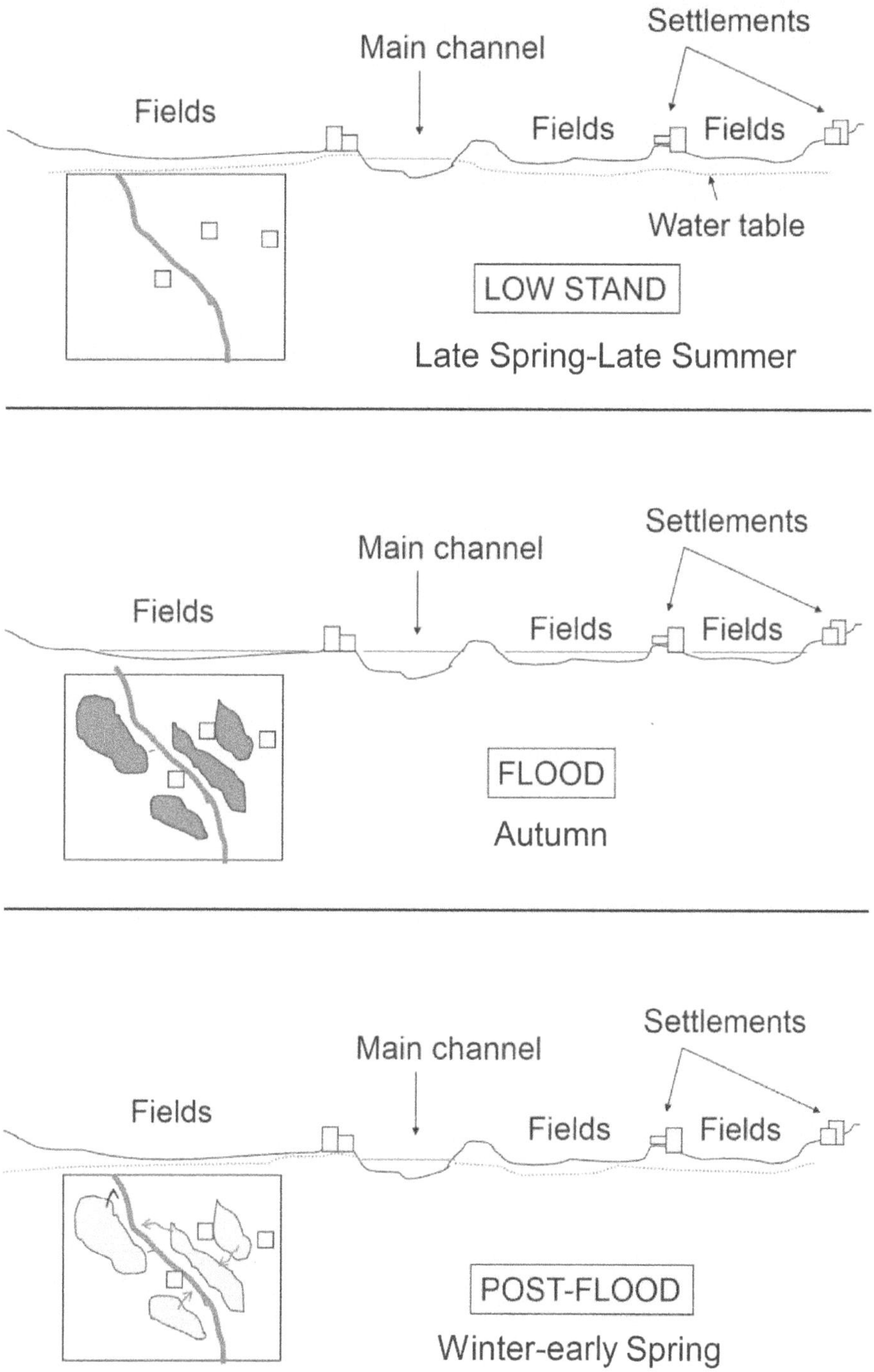

Figure 10.2. Egyptian basin irrigation system. Based on description in Butzer (1976).

Yellow River flood plain made the main channel to shift south emptying into the sea at a different location (Figure 10.2). One of the peaks in sedimentation and avulsion occurred during the Western Han Dynasty (206 BCE–23 CE), during which centralization caused poor management practices (Kidder and Liu 2014), as confirmed through extensive geoarchaeological research, which has revealed evidence of ancient efforts to avert the effects of river avulsion (See Kidder and Liu 2014; Kidder and Zhuang 2014).

Although channel avulsion was common in the Nile Delta, it never had the catastrophic effects it had in China, in part because of the nature of the Nile Valley, draining areas with more stable slopes (non-loess) and less disturbed areas. Like the Mesopotamian rivers, the Yellow River had a relatively diverse and mountainous catchment, with high disturbance at times but with moderately less erodible materials. Vulnerability was high in all cases, as failure in the source of water, the monsoon or the cyclogenesis in the Mediterranean, were beyond control of the governments and farmers, yet the risks were very high because the investments, particularly in the Mesopotamian and Yellow river cases, were high. The Egyptian system required much less in terms of construction and organization, since it was more contained in the narrow Nile Plain.

It is difficult to determine which of the three systems was more sustainable. When comparing the Mesopotamian radial and Egyptian basin systems, the latter seems to be more appropriate in terms of avoiding widespread soil salinization (Butzer 1976). However, this may be just a matter of floodplain morphology, where the water table was for the most part lower in the concave basin of the Nile (Figure 10.1). The Tigris and Euphrates in the lower Mesopotamia had a convex flood plain more prone not only to flooding, but to a high water table conducive to salinization (Adams 1965; Butzer 1976). The sustainability of the Egyptian basin system also implied that the year-round system utilizing different parts of the basin, the canals, and the water of the river made it more sustainable because of its year-round productivity (Butzer 1976).

The three cases show how important it is to understand institutional functionality, centralization, and organization, all of which are aspects that in good or bad times save a civilization (Butzer 2012). The problem is, however, when decentralization becomes dysfunctional either by inefficiency or poor judgment and action (Butzer 2012). Poor judgment partly has to do with risk assessment, an aspect that is difficult to assess by studying only large sites. As shown in the case of the Yellow River, the study of sediments, geomorphology, historic sources, and the overall idea of time and space provides a better look at what were good and bad decisions (See Kidder and Liu 2014).

CASE 10.2: THE CLASSIC MAYA COLLAPSE AND THE DEGRADATION OF SOILS IN THE MAYA LOWLANDS: GEOARCHAEOLOGICAL MODELS OF LANDSCAPE TRANSFORMATION

The Classic Maya collapse

The Prehispanic Mayan history is divided into the Formative (Preclassic), Classic, and Postclassic, of which the Classic period is the one that produced the splendor of the Mayan civilization. During the Classic period (approximately 300–1000 CE) most of the large cities were concentrated in the central part of the southern Maya lowlands (Figure 10.4a). This

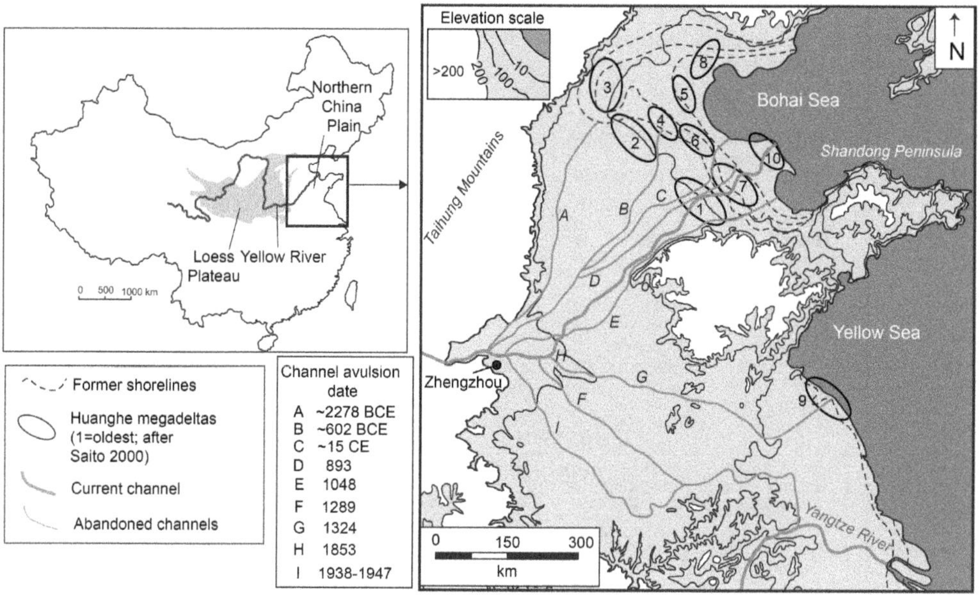

Figure 10.3. a) Location of the Yellow River Basin, the Loess Plateau, and the northern China Plain; b) Avulsion in the lower Yellow River Delta in the Northern China Plain. Modified from Saito (2000) and Kidder and Liu (2014).

was the area with the highest concentration of population in comparison with the northern lowlands and the highlands, and one that led to the transformation of the landscape.

The disintegration of the Classic Maya societal system, which occurred between 850 and 1000 CE, is among one of the most discussed and recently researched cases of civilization collapse. Although evident through archaeological research in many of its material remains, the collapse was for many decades associated with political struggle and warfare among the various Mayan city states. It was not until the 1990s that evidence of climatic causes became evident through the study of lake sediments (Hodell, Curtis, and Brenner 1995). The problem then was how to link a climatic phenomenon to the societal collapse. The answer was not easy to find because it needed first a series of datasets at different scales and from different parts across the Maya area (Figure 10.4a–b).

On one hand data from lake deposits provided information on an array of proxies. Sedimentation rates and pollen data were crucial in showing that agricultural activities increasingly expanded after the Preclassic (or Formative) period, ca. 2000 years before the collapse (Kennett and Beach 2013). During the Early Classic (300–600 CE) the population grew steadily, reaching dramatic growth during the Late Classic (600–800 CE) with a relative improvement in precipitation (Figure 4.2) (Webster et al. 2007; Kennett et al. 2012; Medina-Elizalde et al. 2012). This rapid increase is evident in pollen records that show a dramatic decrease in tropical forest taxa in favor of grasses, herbs, and maize (Turner II and Sabloff 2012). This process was concurrent with soil and sediment erosion and the consequent rapid accumulation of sediments in the lakes, forming the so-called *Maya clay*, which, ironically, occurred not in the Classic, but during the low-density population times of the Preclassic Maya (Anselmetti et al. 2007).

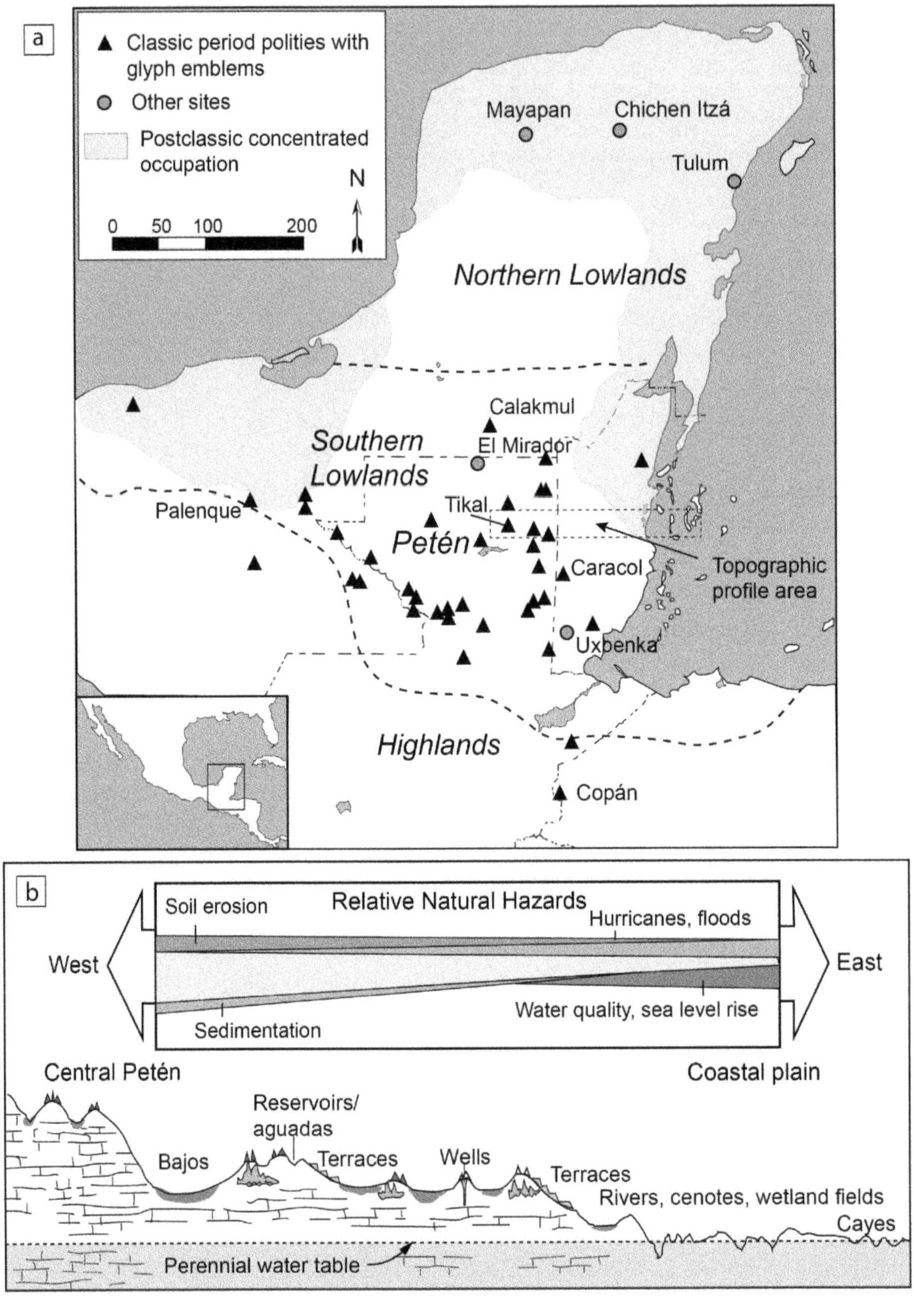

Figure 10.4. a) Map of Classic Maya polities. Modified from Kennett and Beach (2013) with elements from Turner II and Sabloff (2012); and b) hypothesized transect from the central part of the southern lowlands (Petén) to the Caribbean coast (Belize) indicating the relative intensity of the natural hazards in the region. Modified from Dunning, Beach, and Luzzadder-Beach (2012).

The end of the Late Classic (ca. 750 CE) represents the culmination of growth and the slow decline in building construction, coincident with high precipitation variability (Figure 4.2). The situation was aggravated during the Terminal Classic (800–1000 CE), during which variability increased and drought periods were longer. Consequently, at the beginning of the Terminal Classic most cities in the central lowlands were deserted, with only a few remaining modestly occupied a bit longer (e.g., Tikal and Caracol) (Figure 10.4a). A prolonged period of intense drought ensued after 1000 CE, during which the last cities in the central lowlands were deserted. This period coincides with the beginning of the Postclassic, and the reappearance of the Maya culture in the northern part of the Yucatan Peninsula (Dunning and Beach 2010; Turner II and Sabloff 2012). The southern Maya Lowlands area never again experienced the amount of population that occupied the region during the Classic Period.

Although most information on the rise, growth, and collapse of the Classic Maya comes from the archaeological record from the Mayan cities, including historical information obtained from Mayan hieroglyphic inscriptions, little was known about what exactly happened in the areas that produced food, namely the countryside. This, however, changed as geoarchaeological research focused on sediment archives, *landesque* structures, soil structures, and in general processes that occurred in the rural environment.

The countryside, soils, and geomorphic processes

The southern Maya lowlands constitute a hilly, karstic plateau, with concave and convex topographies, and low areas of sediment accumulation and wetlands (*bajos*) (Figure 10.4b). The topography falls towards the Caribbean, with extensive areas of low elevation characterized by flood plains and wetlands. During the Classic period, the wetlands were transformed by draining them with canals to create raised fields (Beach et al. 2008). The cities originally (i.e., in the Early Classic) occupied mainly the hilly areas, with agriculture originally in the lower areas, particularly in the bajos.

As population grew many upland areas were opened for farming through slash-and-burn, with a fallow period continually reduced and exhausting soil nutrients (Dunning and Beach 2010; Turner and Sabloff 2012; Kennett and Beach 2013). Although deforestation during the Classic is evident in pollen data, the rapid transformation of the forest and areas of shifting agriculture into more permanent intensive cropland is evidenced in the isotopic composition of soils. The typical $\delta^{13}C$ -depleted of tropical forest soils would turn into ^{13}C richer soils typical of areas occupied by C_4 grasses. Maize is a C_4 grass as well as many grasses that occupied the opened fallow areas. This change in $\delta^{13}C$ is evident in paleosols buried under the modern forest soils (Figure 10.5a).

As population grew in the Middle and Late Classic, more wetlands were transformed, and new lands for farming opened in low areas were opened as water was diverted into them to create artificial wetlands (Beach et al. 2008, 2009; Dunning, Beach, and Luzzadder-Beach 2012). At the same time more areas on the slopes were cleared, and terraces were built to collect and retain soil (Dunning and Beach 2010). Further evidence of intensification is seen in the removal of soil and accumulation on low areas, which in many cases bury soil (Figure 10.5b).

The stratigraphy of natural wetland sediments and soils would be horizontal, but the desiccation produced by management (drainage) and subsequent drying and alluviation

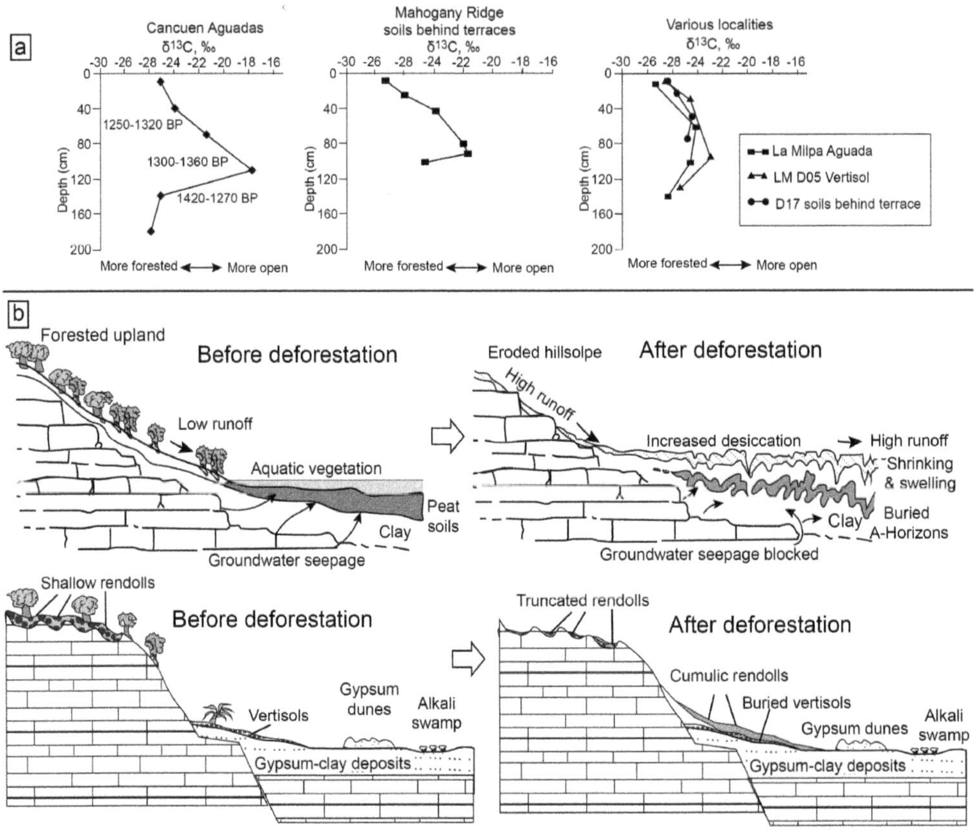

Figure 10.5. a) $\delta^{13}C$ from soil profiles from Beach et al. (2011). b) Models of responses to erosion in uplands and accumulation and soil upwelling in the bajos, after Beach et al. (2009).

caused the shrinking and swelling of the clay-rich soils in the bajos (Figure 10.5b, above). Another persistent feature in soils and deposits in the bajos constitute gypsum removed from the dissolution of the rock and redeposited in the soils (Figure 10.5b, below). Consequently, the amounts of sulfate and chloride resulting from this process contaminate the water and make agriculture in these soils problematic (Beach et al. 2011)

By the time the high variability of precipitation occurred during the Late Classic (600–800 CE) the implemented agricultural systems had begun. Erosion began to remove more soil from the slopes and carried it to the wetlands. The process was exacerbated as some of the region began to be abandoned. Examples of canals filled in with sediments or Classic Maya soils mantled by colluvium and alluvium are a common site in the geoarchaeological record of the bajos (Fig.10.5b) (Dunning, Beach, and Luzzadder-Beach 2012).

Stabilization of geomorphic systems occurred again as forests regrew and soils were regenerated. Proxies from lake sediments and land provide estimates for a relatively rapid regeneration. It took between 80 and 260 years after abandonment for the forest to regenerate and between 120 and 280 years for soils to become stable (Mueller et al. 2010). Interestingly, this process occurred while the area was still under drought, an aspect that shows the natural resilience of forest systems in this part of the tropics.

Ironically, despite widespread soil eriosion in the Late and Terminal Classic, sediment removal was minimal compared to earlier periods. The Maya Clays, identified in the cores of Lake Salpeten, correspond to high sedimentation rates in the Preclassic period, with considerable decrease during the Early and Middle Classic periods (Anselmetti et al. 2007), suggesting that during the later periods soil erosion control measures were implemented (Kennett et al. 2012).

Sustainable systems in the Classic Maya?

One of the main arguments for its collapse was that the Classic Maya built unsustainable systems. However, geoarchaeological research shows that this was not always the case. In fact, the use of the wetlands in the bajos and the construction of terraces, and overall the management of water, was well engineered and relatively sustainable. The problem was that the *landesque* complex system could not cope with the high precipitation variability and drought. Certainly, there were aspects of poor management, particularly through the cultivation of marginal areas of poor, shallow, stony soils occupied by forest and steep slopes, but this was not the general case across the CML.

Some of the cities that survived into the Terminal Classic do show examples of sustainable systems, as is the case of water management around Tikal (Scarborough et al. 2012; Lentz et al. 2014) and the intense terracing of slopes around Caracol, in part due to the terracing system responding to a less centralized population (Chase et al. 2011). In the end, however, these cities were depopulated as the drought intensified after 900 AD (Figure 4.2; Figure 10.4a–b).

In summary, geoarchaeological research of the CML countryside has revealed not only that the Maya did manage the environment in a relatively sustainable way, but also that management was unequal across the region, where different states managed their environment differently. Things went wrong in the end because, despite its sustainability, the agricultural systems were highly vulnerable, particularly in the face of climatic variability and drought.

Although more data has helped explain the collapse of the Classic Maya in the CML, there are more questions for geoarchaeologists to help answer, and one of them was why the Maya Civilization moved to the northern part of the Peninsula, an area that receives less precipitation and has no rivers and few wetlands. Although some economic aspects have been proposed (e.g., maritime trade), the aspect of management of the land for subsistence is important.

11

The Geoarchaeology of Rural Landscapes

INTRODUCTION

Rural landscapes cannot be dissociated from urban landscapes because they are also part of the same complex society and societal-environmental interaction network. However, rural landscapes present formation processes and contextual issues of their own. For this reason, most ancient rural territories are approached at the level of landscape, namely landscape archaeology.

The environmental contextual level presents a series of problems in the interpretation of the record, merely because in highly transformed rural landscapes it is difficult to obtain background environmental information. As seen in the case of the Classic Maya agricultural landscape in the previous chapter, most environmental information comes from paleoenvironmental archives in lake sediments and speleothemes, and less from the agricultural landscapes themselves (Chapter 10, Case 10.1). But that is not always the case, thus most records contain cultural noises (Case 11.1 and 10.2).

In addition to introducing the reader to a series of concepts and approaches to studying rural landscapes in geoarchaeology, this chapter discusses two case studies, one referring to the features created by ancient Greek colonists in Crimea (Case 11.1), and another to the earliest evidence of farming the lacustrine and palustrine environments of the Basin of Mexico through the construction of raised fields (*chinampas*) (Case 11.2). Both cases denote in a way some level of social and environmental complexity and both are examples of extensive research using geoarchaeological methods.

THE RURAL LANDSCAPE: CONCEPTS AND ENVIRONMENTAL APPROACHES

Conceptualizing the rural landscape

Determining the limits between urban and rural realms has been a complicated task even in some modern contexts. Sociologists, demographers, and human geographers have come up with different ways for determining urban limits, sometimes based on land use, population density, and economic activities. In archaeological contexts, stablishing urban boundaries is even more difficult, in part because of the fragmentary nature of the archaeological record, but also because what we consider urban today may have not been so in the context of past cultures.

Where there are highly nucleated settlements, the distinction between urban and rural is more evident, as in the Greco-Roman world, where the distinction between urban and rural realms was planned and emphasized. Such a notion of the division of space was most likely adopted from the Greek notion of *polis* and *chora* (Berque 2009). The chora would have been the area surrounding the urbanized areas (polis). In Medieval Europe, and in ancient Asia, the Near East, North Africa, and Sahelian Africa, urbanization often occurred inside city walls (tells) or compounds of agglomerated buildings that were created for defensive purposes and not for separation of the urban and rural. The latter in some cases meant territories that were not necessarily farmland, but also areas of nomadic pastoralism.

The distinction between urban and rural realms is difficult to determine in some parts of ancient Mesoamerica. Some of the Mayan cities were in fact low-density rural settlements with houses surrounded by orchards and cropland, showing that agriculture was practised within the city (Turner II and Sabloff 2012). The historical references and urban archaeology of Aztec capital, Tenochtitlan, suggest that the majority of its population lived and farmed in artificial islands separated by canals (Calnek 1972). Only the central part of the "city, with its religious, administrative, and commercial centers," would have been more like a city in the Western sense of urbanism.

But despite all the different variations regarding the urban-rural boundary, the important aspect to consider is that agricultural landscapes should be understood principally as areas where foodstuffs are produced through farming and raising animals. Such areas may or may not include the territories with seasonal pastures, forests, lakes, and rivers, some of which may be part of the hinterland of a city or state, but are not always rural, despite interacting with the production of foodstuffs.

Incremental change and landesque capital in the geoarchaeological record

From the point of view of the geoarchaeological record (Table 3.2), the rural landscape contains visible and invisible features. They include natural features, or features with less or no modification by humans, and cultural features, which include infrastructure (canals, dams, terraces, etc.) and human-made soils (i.e., anthrosols). These features are dynamic in the sense of spacetime, that is to say, they may belong to one period, but have been modified through later periods.

Such modifications can be continuous maintenance and upgrading of old structures or repairs and rehabilitation of old abandoned structures. Such modification in rural landscapes are similar to modifications of the urban landscape where buildings are repaired or reused, or in many cases when they are destroyed, their materials are recycled. In the rural landscape this idea of change by adding or removing parts to agricultural features is defined as an *incremental process* in agroecosystems (Doolittle 1984; 2001).

Somewhat different in terms of dynamics, but also important in the transformation of the land, is the concept of *landesque capital,* defined as an investment in land as a way to improve the agricultural potential (Blaike and Brookfield 1987). In essence, such investment encompasses all the structures built and maintained with the purpose of agricultural production. The idea of capital is metaphoric in the sense that the agricultural value of the landscape represents an asset and an investment with payoffs in the future (Morrison 2014). The implication of this idea is that agricultural infrastructure is an asset, like a house or a car, which has to be repaired and maintained to keep a good value in case it has to be sold.

The term *landesque* has drawn criticism, particularly because it replaces the more commonly used term *landscape* (Morrison 2014; Widgren and Håkansson 2014). However, landesque capital entails the idea of labor or investment through labor (Widgren and Håkansson 2014), which is an idea that "landscape capital" may not convey.

Another source of criticism comes from the idea that labor contrasts with technology, as is the case of labor-tasking and techno-tasking, two processes in land development that may be conducive to land productivity, but work in different ways (Scarborough 2003). Also, criticism of the idea of "capital" comes with the idea that some landesque structures may not have economic purposes, but be the result of traditional practices or have symbolic value (Widgren and Håkansson 2014).

But beyond the semantics and generalizations involved in the term, geographers, anthropologists, and archaeologists use the term to convey the idea of improvement of the land, which in many ways means constructing and repairing infrastructure for agriculture, whether it is canals, dams, terraces, or any other structure.

Sometimes, however, such agricultural structures may be part of what William Doolittle (1984) calls *systematic change*, which, as opposed to incremental change, includes modifications that are not meant to endure. Thus, establishing the boundaries between the usable and reusable can be difficult, depending on the culture (e.g., semi-sedentary) or the geomorphic system (e.g., a highly dynamic system). Besides, certain investments are a gamble, and are not necessarily expected to last.

The significance of the *landesque capital* and its criticism are relevant to geoarchaeology in the sense that through geoarchaeological research it is possible to determine the durability of agricultural infrastructure through time. This in part has to do with the geomorphic history of a particular landscape, but also with the spacetime relations between features, some of which can be dated. The case of the Mayan agricultural landscape (Chapter 10, Case 10.2) shows that some structures were not built with the idea of creating an asset, but with the idea of solving an immediate problem, whether it was erosion or water scarcity. Similarly, the irrigation infrastructures can be constructed seasonally, as is the case of the labor investment in the irrigation systems of the Middle Balsas region (Case 16.1, Chapter 16).

Ethnoarchaeological and ethnohistorical approaches

With the growing practice of ethnogeoarchaeology, more and more aspects of rural land management in modern traditional societies are implemented in addition to numerous experiments (e.g., the Butser Farm in England). Likewise, for aspects of management that survived until recent times, historical and ethnohistorical records have become important. However, great caution is needed when using these recent references to interpret the complex societal-environmental systems (CSES) in past times when environmental and social conditions were different.

One interesting aspect of the interpretation of processes in the three fluvial systems discussed here is the reliance on modern analogs, ethnohistoric examples, and historical records. For example, the ancient Mesopotamian radial irrigation system was reconstructed based on the surviving system of radial irrigation in the Diyala River, a tributary of the Tigris (Adams 1965). The system was assumed to be similar along the Tigris and Euphrates, where the system disappeared a long time ago.

Likewise, the case of modern surviving lacustrine raised fields (chinampas) south of Mexico City are often seen as ethnographic modern references to Prehispanic systems of chinampas. However, it has been pointed out that archaeological chinampas had different dimensions and structure (Frederick 2007) and were real means of subsistence agriculture (Luna-Golya 2014). In contrast, modern chinampas operate under modern hydrological and economic contexts that require labor and techniques different from their Prehispanic counterparts (Crossley 2004; Luna-Golya 2014).

GEOARCHAEOLOGICAL STRATEGIES IN ANCIENT RURAL CONTEXTS

The view from above and below

Visible cultural features on the surface are commonly mapped through a combination of pedestrian survey and remote sensing. Features that are not visible but are tangible are mapped either through non-invasive methods (i.e., magnetometry and GPR), or through coring and test pit digging. Features are displayed spatially in maps, often organized in time sequences or periods, thus providing a rough idea of change. The resulting maps, however, provide little information on formation processes, which are better expressed in the stratigraphy.

The stratigraphy provides a view to C and N transform processes through the array of soils and sediments, are significant features in agricultural landscapes that are not recorded through surveys. Traces of past agriculture, or other activities, are distinguished from soil structure, soil microstratigraphy, and chemical traces. From another view, soils are also important sources of information related to their formation and to productivity.

The combination from maps, soils and sediments, and topography, help reconstruct the geomorphic history of the landscape, which is one of the main objectives of geoarchaeological research. Such history can be broken down into snapshots that could be used as schematic models representing paleolandscapes at different times (e.g., Figure 4.1).

Mixtures of artifacts and features in rural contexts

Agricultural landscapes are used and abandoned, re-used and abandoned again. Innovations are sometimes introduced, some of which may obliterate the remains of previous periods, reuse them, or superimpose new features. In other words, agricultural landscapes are essentially palimpsests where there are features (built structures and soils) of several periods, a typical situation in areas with continuous rural use through several periods.

Dating agricultural terraces and dams using artifacts in areas with continuous or intermittent rural use could be a problem because of multiple phases of recycling of soils and sediments, and construction materials, through continuous repairs and reclamations.

This results in soils and deposits with mixtures of artifacts of different periods, making archaeological dating difficult. Likewise, datable organic material might be affected by processes of land reclamation and recycling of organic soils.

Agricultural terraces that contain artifacts from several periods are common throughout many parts of the Americas, particularly central Mexico and the Andes, where reclamation was an important part of reusing previously abandoned terraces (Donkin 1979). In the Andes, cases of multiple repairs by archaeological, historical, and modern cultures have kept terraces functioning to this day (Goodman-Elgar 2008a). Similarly, cases of repair are documented in some parts of the Basin of Mexico (Cordova 1997).

In the Mediterranean region, there are multiple cases of terraces that have been reused. One case of many that illustrates the process has been studied in hills west and south of Jerusalem. Archaeological and numerical dating of structures and sediments has shown that some of these terraces, which were still in use in the early twentieth century, date back to the early Byzantine period (Davidovich et al. 2012), and some to the Early Bronze Age (Gibson, Ibbs, and Kloner 1991). The millennial use of this terracing system was possible through multiple repairs, even if some terraces were abandoned for some time. Even today in that area, terraces abandoned as a result of the population change in the war of 1947–1948 that have shown some damage need only minor repair to become functional again (Figure 11.1).

Although multiple construction phases are encountered, sometimes detailed geoarchaeological study can reveal minor details that can help date agricultural features to a particular period. But if that is not possible, sometimes using the oldest and youngest possible date can help place certain features on a time frame. The importance of properly dating landesque infrastructure terraces lies in providing a time frame for the initial construction of features and for the times of use and abandonment.

CASE 11.1: THE ANCIENT GREEK RURAL LANDSCAPE IN SOUTHWESTERN CRIMEA

Historic background

Population growth in the Greek mainland during the Greek Archaic (800–480 BCE) resulted in the spread of colonies along the Mediterranean shores, some of which formed city states or alliances of ethnic states. By the sixth century BCE, Greek colonies began to proliferate along the coasts of the Black Sea, occupying places as far north as the mouths of the rivers Dnieper and Don.

In this major process of Greek migration and colonization, the Crimean Peninsula became the center of at least two major political entities, Panticapaion (near present-day Kerch) and Tauric Chersonesos (in present-day Sevastopol), in the southwest (Figure 11.2a). Both cities had their allied smaller towns and their respective agricultural territories. The main focus here is the one established near Tauric Chersonesos, which was founded in the late sixth century BC by Doric settlers from Heraklea Pontika (in present-day Ereğli in northern Turkey).

In the subsequent centuries, Tauric Chersonesos became an important agricultural and commercial center dedicated to the export of fish, wine, and grain (Carter et al. 2000). Its agricultural territory was divided into an outer chora that extended along the west coast of Crimea, including large parts of the Tarkhankut Peninsula (Figure 11.2a), and an immediate

Figure 11.1. View of terraces in the Sataf district east of Jerusalem. Some of these terraces date back several millennia. Photographed by the author.

chora located in the Heraklean Peninsula around the polis, or city, of Tauric Chersonesos (Figure 11.2b).

The territory of the outer chora has been extensively surveyed, producing a great number of farmsites, division lines, vineyard rows, and other visible features (Stolba and Andressen 2014; Smekalova et al. 2016) as well as biochemical traces of past agriculture in

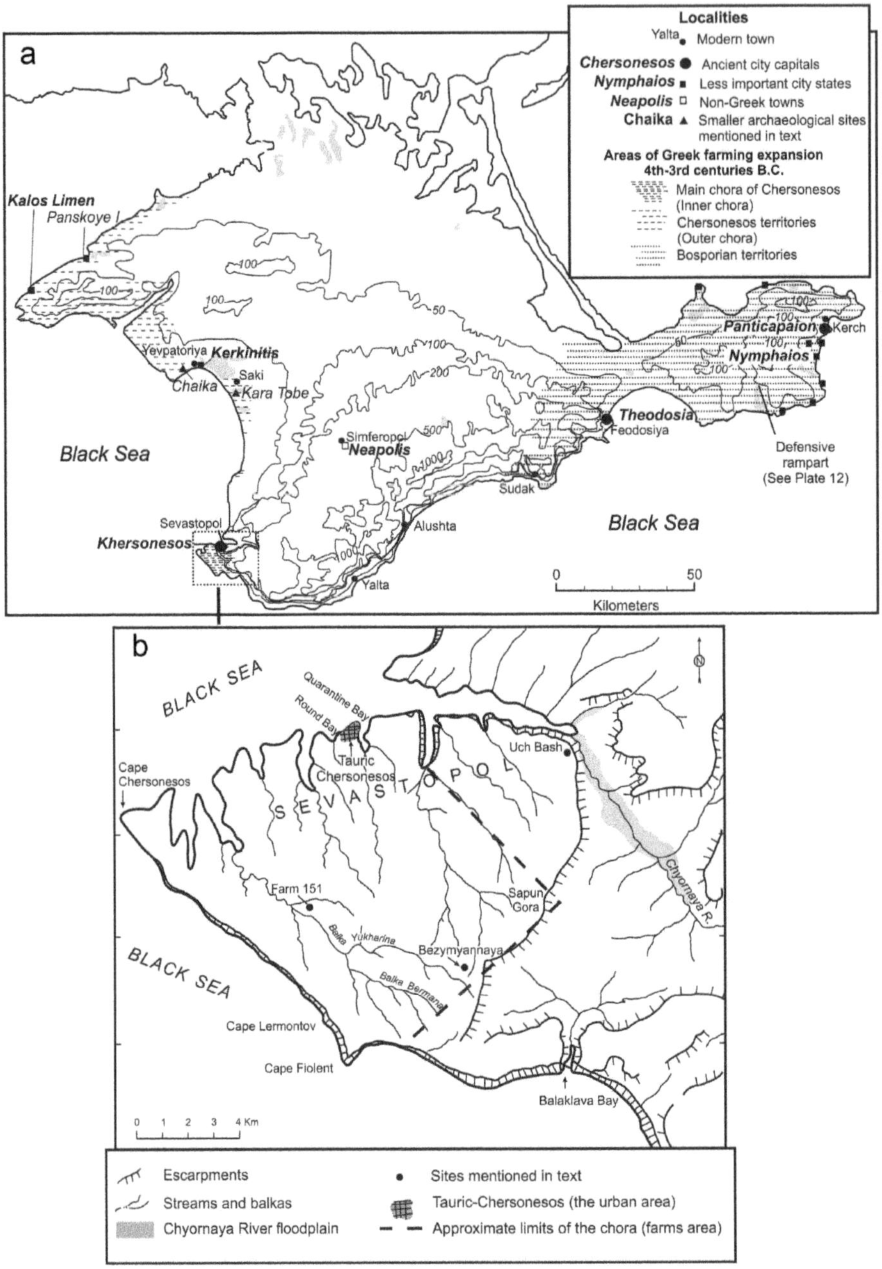

Figure 11.2. (a) Crimea and the ancient Greek farming territories (second half of the first millennium BCE) showing the Tauric Chersonesos territories in the west; (b) The Heraklean Peninsula with the city of Tauric Chersonesos and its immediate chora.

soils (Lisetskii, Smekalova, and Marinina 2016). This territory, however, had a much shorter occupation than that of the immediate chora, as it was abandoned in the second century. The causes of its abandonment have been debated, ranging from invasions from peoples in the steppes to warfare with the Bosporan Kingdom, established in Panticapaion in the east, and climatic deterioration (Cordova et al. 2011).

The territory of the immediate chora (Figure 11.2b), on the other hand, has more permanent structures and well-delineated farms, terraces, dams, aqueducts (i.e., *landesque* infrastructure) and defensive towers. This area also lies in a wetter and warmer part of the Crimean Peninsula and close to the city. Nonetheless, an apparent agricultural decline in the immediate chora occurred slowly through the Roman and Byzantine periods. The city, however, did not disappear, as it continued to be occupied until its destruction by the Tatar-Mongols in 1399. Although the city played an important role as a Byzantine hub, land use in the countryside was different, a change that is evident in the pollen records (Cordova and Lehman 2003). Agricultural activity shifted, probably to other areas beyond the chora and most likely to the mountain valleys (Cordova 2016).

Yet, there are several questions regarding the environmental aspects of the original Greek colonization and layout of features in the chora from its establishment to its decline. Therefore, a strategy to study paleoenvironments and paleolandscapes requires a strategy to study sediments and proxies obtained from them.

Some studies of the land and in particular the farms have been carried out since the 1930s, particularly on aspects of the division of land and walls retaining sediments (see references in Cordova and Lehman 2005; Cordova et al. 2011; and Cordova 2008). Studies of plant remains have added some relevant information regarding crops, but nothing about vegetation surroundings. Much less is known about landscape and environmental change before, during, and after the Greek colonization. Thus, a geoarchaeological and paleoecological study is needed to understand the landscapes found by Greek colonists and their subsequent transformation.

Geoarchaeological strategy

Geoarchaeological work has been carried out in conjunction with excavation and restoration of two farm sites, Farm 151 and Bezimiannaya (Figure 11.2b). Most information on sediments and soils came from the deposits inside ravines locally known as *balki*. Two of the balki that provided relevant information are Balka Yukharina and Balka Bermana (Figure 11.2b). The balki in the Heraklean Peninsula are relict fluvial valleys with no running water at present (Cordova et al. 2011). Thus, the balki have been depositories of sediments, some of which contain soils.

The strategy also included the study of sediments of the Chyornaya River flood plain, which although outside the chora, provided an off-site setting. The Chyornaya flood plain coring revealed recent alluvial deposits capping palustrine and lacustrine deposits containing pollen. Further research has been accomplished in other areas of the mountains beyond the chora as a way to obtain regional patterns of change.

At Farm 151 a survey of the immediate surroundings revealed interesting information on agricultural structures, including ancient vineyard rows (Figure 11.3). The nearby Balka Yukharina contains a sequence of sediments that provides information on landscape change from the early Holocene to about the time of the Greek colonization. The top of the sequence had a soil farmed during the Greek period and possibly also during the Byzantine period.

Figure 11.3. (a) Aerial view of Farm 151 indicating the location of Balka Yukharina and other features of the ancient rural landscape; (b) stone rows of ancient vineyards associated with Farm 151; (c) one of the stratigraphic sequences in Balka Yukharina (section AA). Modified from Cordova (2016).

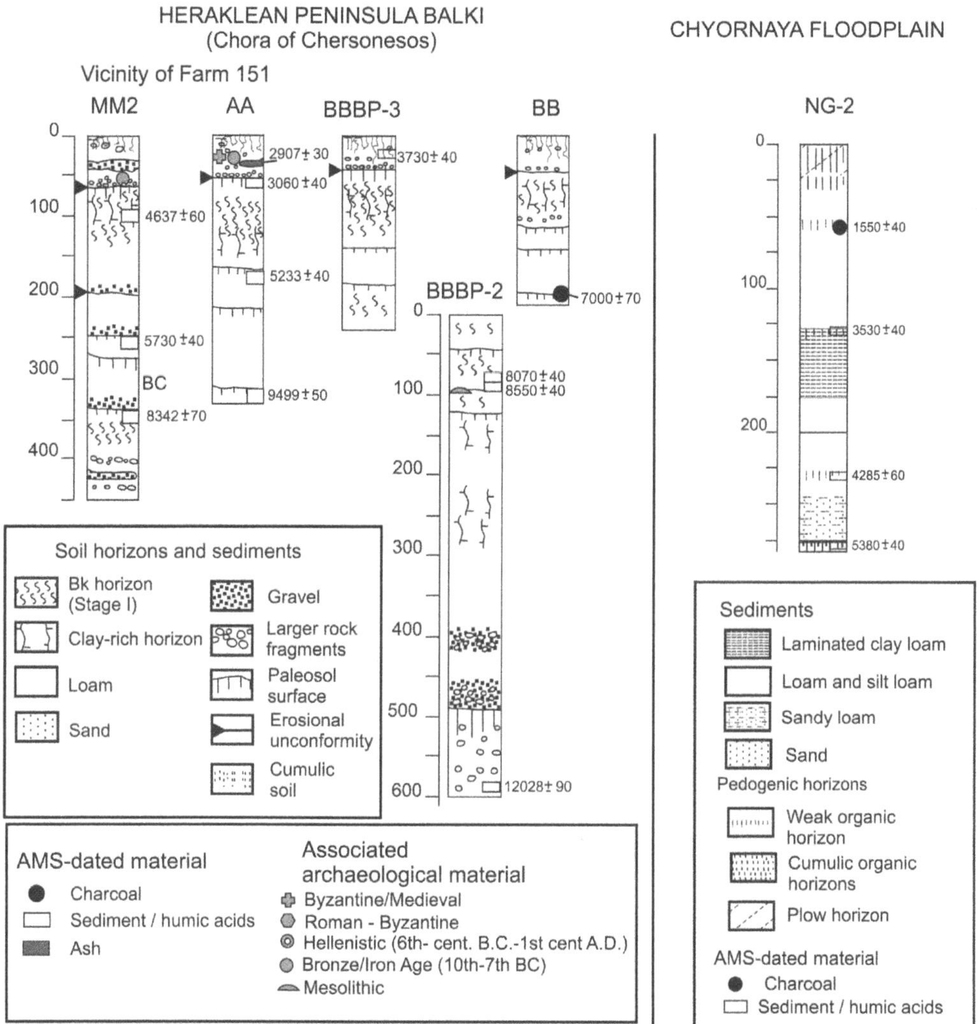

Figure 11.4. Stratigraphic relations between localities in the chora of Chersonesos and the Chyornaya Floodplain. After Cordova et al. (2011).

Stratigraphic sequences and landscape change

The stratigraphic sequences in the balki and the Chyornaya River flood plain provide a sequence of change, which combines sedimentary and pedogenic changes, suggesting stable and unstable periods (Figure 11.4). Some of these sequences have materials spanning from the Mesolithic to the Roman-Byzantine period. The Chyornaya pollen sequence (Core NG-2) provides information on vegetation change, particularly showing the transformation of the landscape by the establishment of Greek farms (beginning in the fifth century BCE). This change is clear not only in the reduction of tree cover, particularly oak, but also in the presence of several weed taxa, including some associated with eroded surfaces (Cordova and Lehman 2005).

Landscape and environment during the ancient Greek colonization

The overall results of the geoarchaeological study in the Heraklean Peninsula and adjacent areas provide a picture of Greek colonization. First, the area already had thin calcareous soils, similar to those of the Mediterranean, for which it was necessary to construct beds, using sediments from the balki. Second, the landscape of the Peninsula was more forested, although it was not forest, but perhaps a forest-steppe with some native Mediterranean elements (Cordova et al. 2005).

The decline of farming is also clear in the pollen records, as woody vegetation continued, but crops and land use changed. The sedimentation in the Chyornaya and other locales in the valleys of the mountains also indicate erosion on the slopes, probably associated with the establishment of farms and cities and pastoral activities during the Middle Ages, suggesting the clearance of forests by the increase of population resulting from the Great Migration period (third to eighth century CE) (Cordova 2016a).

Finally, it is important to stress that the geoarchaeological study also included landscape changes after the Greek colonization, particularly during the unprecedented increase in settled areas in the Middle Ages and in the past two centuries. In particular, the establishment of the fleet base in Sevastopol, and the destructive effect of battles of two wars, the Crimean War and World War II, as well as changes during the Cold War period (see Case 17.1, Chapter 17) had a significant impact on the archaeological features and the geoarchaeological record. This shows the importance of studying recent landscape changes when studying ancient features and landscapes.

CASE 11.2: XALTOCAN: THE GEOARCHAEOLOGY OF A COMPLEX LACUSTRINE SOCIETY BEFORE TENOCHTITLAN

The raised field (chinampa) cultivation in the Basin of Mexico

Spanish conquistadors and chroniclers marveled at the magnificent appearance and functioning of Tenochtitlan, the Aztec capital, a city totally adapted to a lacustrine environment, and functioning as both and urban center and agricultural territory. Its success as an agricultural territory was based on the creation of raised fields locally known as *chinampas*. Located in many parts of the lakes of the Basin of Mexico, the chinampa is a cultivation system that consists in the creation of a soil bed with organic-rich mud from the bottom of a lake, creating at the same time a canal and a bed (Frederick 2007). The canals are used as drainage, but also for reaching the fields by canoe. Today, the chinampa system survives in what is left of Lake Xochimilco (Figure 11.5).

In the sixteenth century, the Basin of Mexico was an endorheic basin whose waters were collected in five interconnected lakes (Figure 11.5). The southern lakes (Xochimilco and Chalco) were fed by springs and rivers originating from the forested and more humid parts of the southern mountains, while the other lakes received waters from rivers draining areas rocks that added salt minerals to its waters. Consequently, the waters

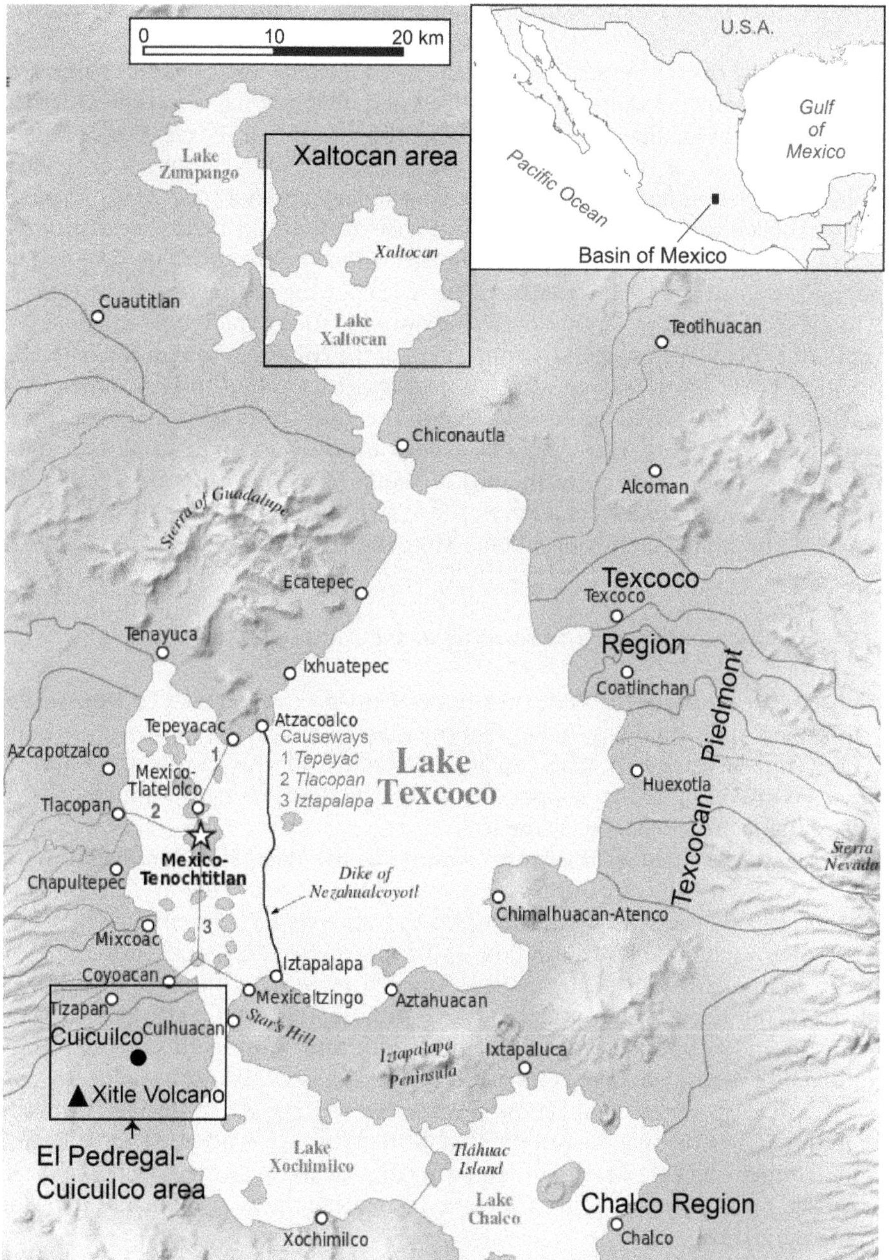

Figure 11.5. The lakes of the Basin of Mexico with the location of Xaltocan in the north (Case 11.2), the area of Xitle volcano and Cuicuilco (Case 13.1) and the Texcoco Region (Case 15.1).

of the other three lakes (Texococo, Xaltocan, and Zumpango) were brackish and at times saline.

The lakes of the basin were shallow and subject to seasonal and sporadic changes causing sometimes flood, which required building dikes and diverting rivers to protect settlements (Candiani 2014). The shallow nature of the lakes also meant that the creation of raised fields would not be difficult, as long as the water was not saline and the sediment excavated from the canals was rich in nutrients (mainly organic sediment, and not clay). This was possible in the southern lakes and in the eastern half of Lake Texcoco (i.e., Lake Mexico) under the construction of the dikes that protected the area from the incursion of salty water from the east and retained the fresh water from the rivers flowing in from the west (Figure 11.5).

Archaeological data from various localities from what used to be Tenochtitlan and various parts of the southern Basin suggest that the origin of the chinampa system is relatively recent (approximately late fourteenth to fifteenth century). Its location in fresh-water lakes also suggests a strategy that could not have existed in the more saline lakes Texcoco, Zumpango, and Xaltocan (Figure 11.5). However, archaeological survey and excavation in conjunction with geoarchaeological research in the bed of the former Lake Xaltocan revealed that the chinampa system at this locality was developed at an earlier time than in the rest of the basin (Frederick, Winsborough, and Popper 2005; Morehart 2012b; Morehart and Frederick 2014).

Geoarchaeological research at the Xaltocan chinampas

Xaltocan was an insular settlement on the northern part of the lake by the same name. Geoarchaeological research has shown that the island is an artificial promontory built with sand from the lacustrine beaches and paleobeaches to the northeast (Figure 11.5a) (Frederick, Winsborough, and Popper 2005). The modern town of Xaltocan sits on the remains of the archaeological settlement.

The town was founded in the tenth century, at the time Tula dominated, by Otomi settlers. They deserted the city in the thirteenth century after a war with Cuautitlan (to the west of the lake) and the town was re-occupied by Nahua settlers (Brumfiel 2005), who made it the capital of a state that lasted until the end of the fourteenth century, when the city was conquered by the kingdom of Azcapotzalco, which in turn, in the fifteenth century was conquered by the Triple Alliance (i.e., the Aztec Empire) (Brumfiel 2005; Morehart 2012b).

During its independent years in the thirteenth and fourteenth centuries the city's agricultural base depended on the chinampa system around it. To maintain the system, and particularly to keep it from salinization, fresh water was brought in through canals from a spring in the Cerro Chiconautla and the Cuauhtitlan River (see Figure 11.5). The decline of the system during the Late Aztec period may have occurred as changes in the hydrology of the basin and terracing in the Cerro Chiconautla reduced the flow of fresh water (Morehart and Frederick 2014).

Aerial photography, off-site research, and excavation of the lake bed has provided the geomorphic context of the Xaltocan settlement and its chinampa system (Figure 11.6a). To the northwest of the Xaltocan insular settlement lies a low-topography area with series of surfaces including paleo-lacustrine terraces and sandy paleobeach landforms. Among other findings, off-site geoarchaeological research revealed that sand from the paleobeaches was used to create the island where the town of Xaltocan sits (Frederick, Winsborough, and Popper 2005).

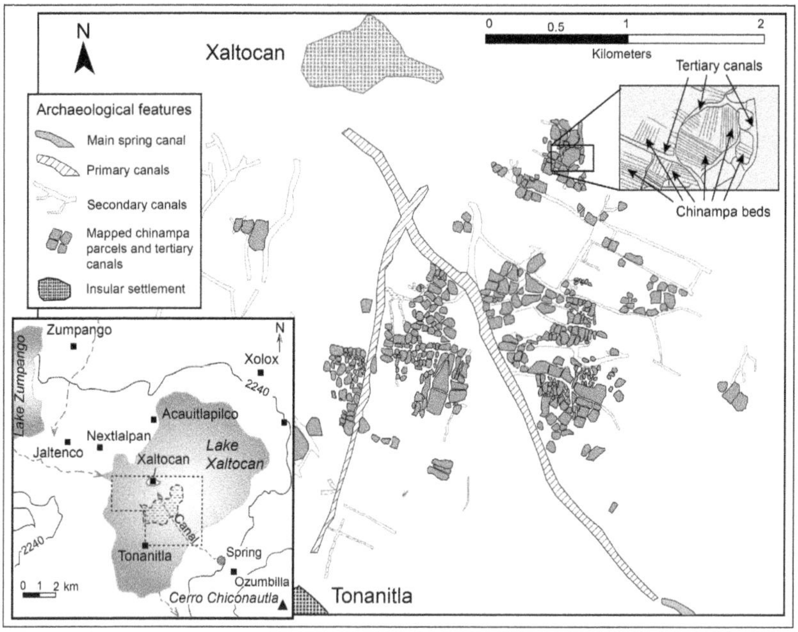

Figure 11.6. a) Geomorphic context of the area (after Frederick, Winsborough, and Popper 2005), and b) mapped system of canals and chinampa beds in the bed of Lake Xaltocan. Modified from Morehart and Frederick (2014).

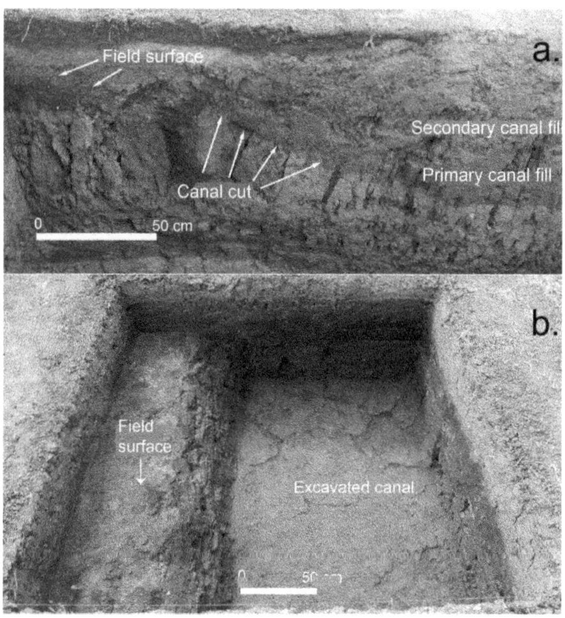

Figure 11.7. Sections of excavated canals in the chinampa fields. Modified from Morehart and Frederick (2014).

The layout of the canals in relation to the chinampas and the insular settlement of Xaltocan provides information on the function and hierarchy of the canals, particularly in terms of drainage and feeding of fresh spring water to the system (Figure 11.6b). The microstratigraphy of the chinampas also allows a careful selection for dating, as it is important to understand the process of sedimentation, the digging of the canals, and the creation of the beds (Figure 11.7).

The combination of remote sensing techniques, GIS, and microstratigraphy show a palimpsest of features in the fields and canals (Morehart 2012a). Radiocarbon assays in combination with archaeological material help date the features, suggesting that the booming phase of the chinampa system coincided with the time of apogee of the city (thirteenth to fourteenth centuries), showing that after the incorporation of the city into the Triple Alliance, the system was rebuilt but never reached the extent and sustainable levels that had before (Morehart 2012b; Morehart and Frederick 2014).

At its maximum, the system of chinampas covered at least 1500–2000 ha (Morehart 2012a), providing more than enough food for the subsistence of the town's residents, who also subsisted on fishing and trade of stone implements that they manufactured (Morehart 2012b). More important is that dates show that the Xaltocan chinampa system predates all the other systems in the Basin of Mexico, which suggests the possibility that later chinampa systems in other lacustrine areas of the Basin (e.g., Tenochtitlan and the southern lakes) may have been modeled after the sustainable Xaltocan chinampa system (Morehart and Frederick 2014).

12

Human-Environmental Approaches to Soils and Paleosols

INTRODUCTION

Soils and pedogenic features are key elements in the interpretation of the geoarchaeological record at the levels of site, setting, landscape, and environment. Although pedology provides the basic tools for interpreting soil formation processes in the record, the study of soils in geoarchaeology requires also the understanding of the human factor in soil formation (Amundson and Jenny 1991; Holliday 2004).

The human factor in soil formation can vary from direct intervention by plowing or deliberately creating soils to more indirect forms involving non-farming practices such as pastoralism, introduction or extirpation of fauna, and the deliberate burning of vegetation. An even more indirect modification of soils, though difficult to detect, occurs through the modification of biochemical cycles as in the atmospheric carbon cycle since the beginning of agriculture (Holliday 2004; Ruddiman 2014).

Considering the importance of human influence in pedogenic processes, soils are not only archives of climatic or geomorphic changes, but also of human-environmental history. However, the idea of soils reflecting human-environmental interactions is sometimes neglected or not properly addressed in geoarchaeology. For this reason, this chapter focuses more on the cultural aspects of soils, not only in terms of soil formation, but in terms of cultural value. Another aspect of soils, regardless of the human influence, is seen in paleosols, a subject of great significance in the geoarchaeological record. For this reason, the concept of paleosol deserves here a discussion particularly in relation to human-environmental relations.

Two selected case studies illustrate the combination of the focus topics described above. One case study discusses the interpretation of European dark earths as an example of the formation of anthrosols with examples from Brussels, Belgium (Case 12.1), focusing on the importance of natural and cultural transformation processes in the formation of soils and their eventual burial. The second case study provides an example of the indirect influences of hunter-gatherers in the formation of prairie soils in the tallgrass prairie of North America (Case 12.2), an aspect that is at the stage of hypothesis but that reinforces the idea that soils that are often seen as natural have in fact been in part shaped by indirect human influences.

THEMATIC APPROACHES TO PALEOSOLS

Definition and varieties

Paleosol is the general term used to refer to ancient soils that are no longer acting as soils, or, in a geological sense, a fossil soil. An alternative term is *geosol*, which has been proposed by the North American Commission on Stratigraphic Nomenclature (NACOSN) and has a more geological connotation (Cremeens and Hart 1995; Retallack 2001). Nonetheless, the term paleosol is the more commonly used in Quaternary geology, paleoecology, and geoarchaeology.

In its general conception as soils not being affected by current pedogenic processes, paleosols are generally classified as *buried*, if they are sealed by a sediment cover; *relict*, if they were formed in the past but not buried; or *exhumed*, if they were buried and then exposed by erosion (Figure 12.1a). However, it has been argued that buried soils constitute the only true paleosols, since they are unaffected by modern pedogenesis (Catt 1998) unless they are not buried deeply enough (Holliday 2004). Modern processes such as roots and animals, illuviation of minerals, and mineralization caused by hydrological processes (water table changes), and diagenesis can still modify the original conditions of paleosols (Catt 1998).

Normally, paleosols, whether buried, exhumed or relict, are studied using the taxonomy and horizons classification of modern soils (Birkeland 1999). However, this is sometimes difficult, unless some modifications are made either by adding terms and modifiers to the existing classification for modern soils (Nettleton, Olson, and Wysocki 2000) or by adding lower-case letters to the typical major horizons A, E B, and C (Reuter 2000).

Paleosols provide significant information for paleolandscape and paleoenvironmental reconstruction, which is why it they are an important tool in geoarchaeology. This importance is manifold, because paleosols could be interpreted in several environmental contexts (e.g., geomorphic, climatic, paleoecological, and cultural), which is an aspect that deserves some distinction and discussion.

Paleosols as geomorphic and sedimentary dynamic structures

Paleosols represent stages in the evolution of the landscape and indicators of the paucity of sedimentation and erosion processes. Thus, whether buried, exhumed, or relict, paleosols are an important tool in reconstructing past geomorphic and sedimentary processes (Birkeland 1999; Retallack 2010). The balance between erosion and sedimentation represents an important tool for determining stability or instability in paleolandscapes and landscape (Figure 12.1b).

The geomorphic stability aspect is of great importance for geoarchaeology because of its implications in the reconstruction of paleoclimatic, and in general of environmental, variables (see sections below). As part of the geoarchaeological record, paleosols represent a visible, and at times invisible, element that has important implications at all contextual levels (Figure 3.2).

The diagrammatic model in Figure 12.1b represents the spectrum of geomorphic stability in relation to erosion and sedimentation as two variables. The central part of the

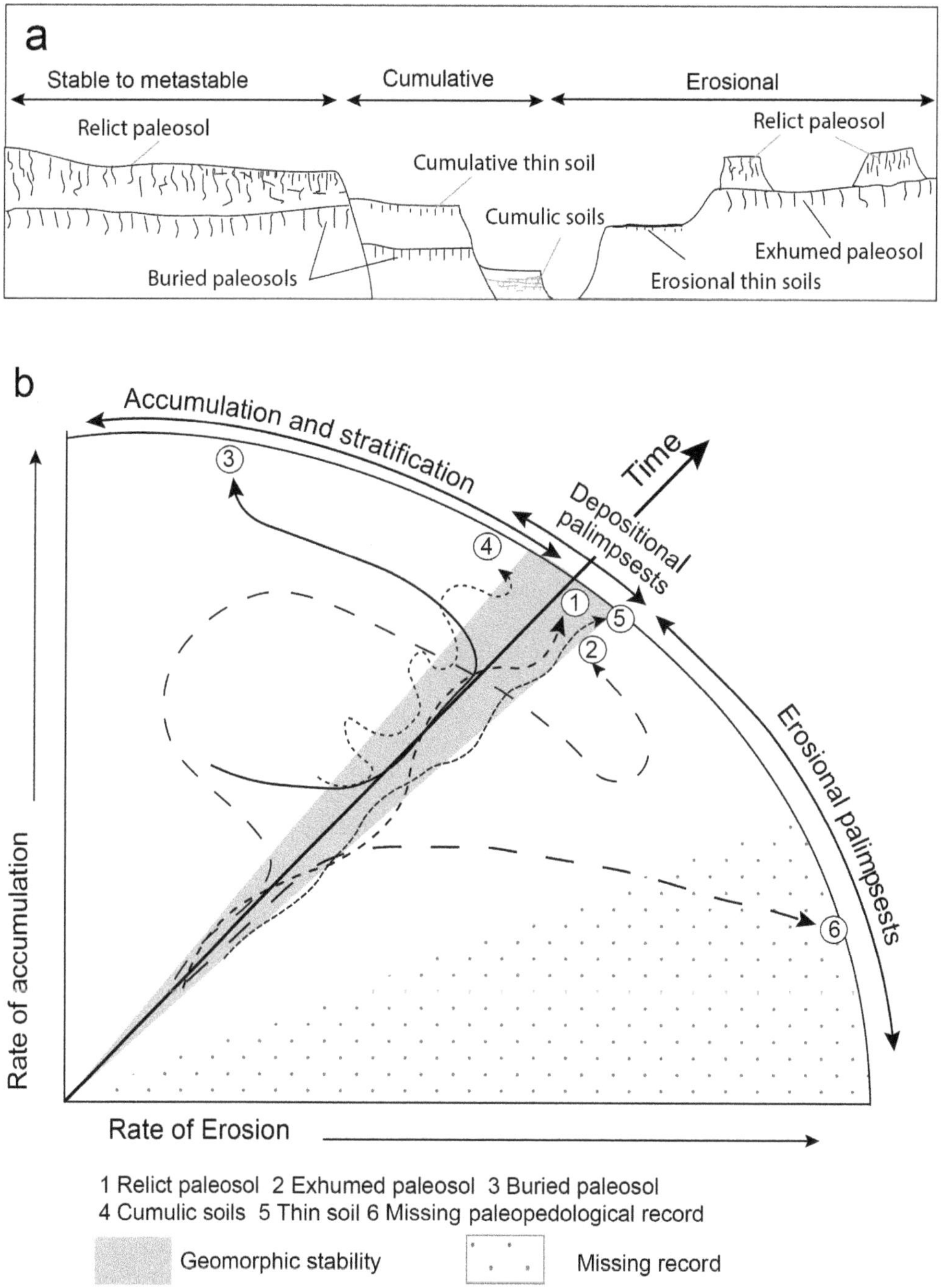

Figure 12.1. a) Idealized geomorphic and stratigraphic setting of different types of paleosols; and b) schematic diagram of paleosol and soil formation and preservation in relation to erosion, sedimentation, time, and landscape stability.

diagram suggests a different level of change, changes in which soils become buried or eroded. Accordingly, geoarchaeologists interpret the evolution of the landscape, say in an alluvial context, by determining phases of erosion, burial, or pedogenic development (Mandel and Holliday 2017).

One interesting aspect of buried paleosols that needs to be born in mind is that humans also create paleosols by deliberately sealing them with structures and sediment or accidentally by catalyzing sedimentation. The record of paleosols below kurgans (burial mounds) in the Eurasian steppes is an important source of information in a landscape where sedimentation of soils is rare (Mitusov et al. 2009). Also, in urban environments where construction overruns rural or even wild areas, soils are buried under the constructions, thus creating a series of records of human-environmental interactions in rural or suburban environments (see Case 12.1).

Paleosols as paleoclimatic archives

In addition to being stratigraphic markers, soils contain information on past climates, obtained not only from their chronological position, but also from their physical and chemical properties (Birkeland 1999). Such properties, when analyzed in the geomorphic context of paleosols, can render information about the climatic conditions that contributed to the formation of that soil.

Additionally, paleosols contain chemical traces that also help reconstruct climate. Stable isotopes from organic matter or carbonates provide information on past climatic conditions. Macro- and micro-plant remains can also add some information obtained through the climatic adaptations of plants sustained by the soil. Macro-remains of plants can be recovered from *histosols*, and in general from soils where organic preservation is good.

Pollen grains are not preserved in most soils, but where they are, they can add valuable information about the vegetation growing on the soil and in its vicinity. Phytoliths are much better preserved in soils than pollen, although they can only reflect part of the vegetation, particularly those plants that produce diagnostic silica bodies. Grass silica phytoliths can also provide information on climatic conditions, by the differentiation of C_3 and C_4 grasses, as well as C_4 mesic and drought-adapted grasses (Mandel and Holliday 2016).

Paleosols as indicators of paleolandscape stability

At the level of landscape, particularly landscape evolution, soils and paleosols are a valuable reference for identifying periods of stability (Holliday, Mandel, and Beach 2016). In theory, a soil is formed where deposition and erosion are at an equilibrium (Figure 12.1b). If erosion exceeds accumulation and pedogenesis, soils will be recognized as either erosive surfaces on old horizons or as poorly developed soils on an erosive surface. If sediment accumulation rates exceed pedogenesis and erosion, then it is possible to have the so-called *cumulic soils* (Holliday 2004). This term is not to be confused with the cumulic horizon of certain soils (see Bockenheim 2014)

However, the model described above can become complicated since the evolution of the landscape itself can change the position of the soil and move it across the accumulation-pedogenesis-erosion spectrum (arrow in Figure 12.1a), thus locking a soil as a paleosol, exposing it to the surface, or eroding it.

Paleosols as paleolandscapes

In soil geomorphology the relations between the landscape and soils are expressed in the soil *catenas* and *soilscapes* (Holliday 2004). Soil catenas are horizontal sequences of soil profiles across different slopes that are used to understand soil formation processes across the landscape. Unlike catenas, soilscapes constitute a more dimensional view of soil formation processes across the landscape. Basically, soilscapes are formed by several catenas.

The concept of soilscape suggests the variability of soils across the landscape; transposed to the concept of paleolandscape, it indicates the diversity of soils that can co-occur during a certain time span. This is an essential aspect to bear in mind because normally our interpretation of the landscape uses a very limited number of soil profiles, which may not be enough to reconstruct the landscape. In this sense, paleosols are also useful for reconstructing paleosurfaces in paleotopographic reconstructions of landscapes buried under sediments from massive flood events, eolian sand and silt, volcanic ashes, and lava.

Paleosols as human-environmental archives

The paleoenvironmental information that can be obtained from paleosols is not all necessarily about climate and other natural factors. Very often geoarchaeologists study paleosols studied in the context of human habitation and use. In many situations paleosols bear artifacts and other evidence of human occupation, and sometimes even evidence of plowing and cultivation.

Numerous proxies obtained from paleosols such as chemical or physical properties as well as microfossils, and micromorphology sections, are used as means of reconstructing agricultural practices. Human influence on paleosols varies, from cases of almost fully human-made soils to cases when humans influence soil formation only slightly.

NATURAL AND ANTHROPIC SPECTRA IN SOIL FORMATION

The human factor in soil formation

Most pedology textbooks state that the factors in soil formation are time, climate, topography, parent material, and organisms (biota). As a conceptual basis this is true, but in the context of human influence on the environment, call it the Anthropocene, soil formation processes are one way or another influenced directly or indirectly by human activities.

Humans influence soils in a variety of ways: as catalysts on the weathering system, through physical and chemical processes, and by diminishing or increasing the effects of other factors of soil formation (e.g., climate, topography, and organisms) (Holliday 2004). For example, topography can be overcome by various methods, and many areas of high

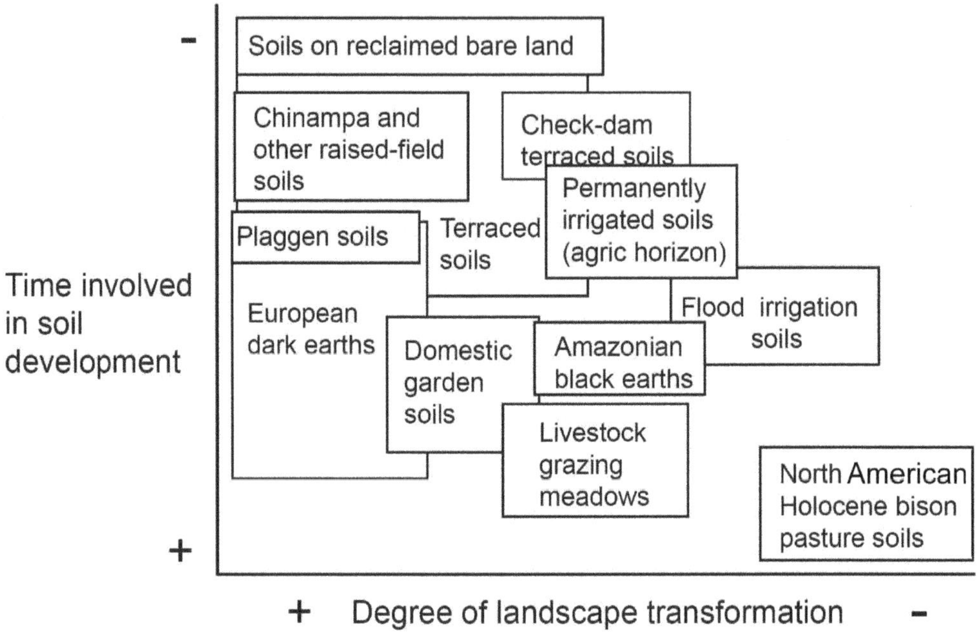

Figure 12.2. Schematic representation of the spectrum of anthropic soils mentioned in text in relation to the degree of landscape transformation (horizontal axis) and time of their formation (vertical axis).

steep slopes that had terracing in the past contain soils that are formed under the influence of humans.

Cases of pristine or unplowed soil, or soils with little human influence, do exist, but they are rare in many parts of the world. Even areas considered pristine ecosystems, such as the Amazonian tropical rainforest, have soils that bear the mark of slash-and-burn agriculture in the past (e.g., *terra preta*) (Denevan 1998). However, where unplowed or unmodified soils exist, they can act as a good reference to study soils that are deeply affected by human activities.

In view of the human influences one can see the degree of anthropogenic influence in soils along a spectrum from more pristine soils with slight human modification to plainly artificial soils, passing through a spectrum of different human influences (Figure 12.2). Most soils created by humans, deliberately or not, are referred to as *anthrosols*, a term that generally means anthropogenic soils.

The broad variety of anthrosols

The spectrum of anthropogenic soils shows the degree of human influence and time involved in their formation (Figure 12.2). Those that are built intentionally develop faster than those that are developed as a result of a cyclic activity or through more passive or unintentional methods. The fast-developing anthropogenic soils include soils built behind terraces, the European plaggen soils, and those of raised fields. At the other end of the spectrum, the slow-developing soils include those formed by slash-burn and fallow cycles or unintentional use under pasture use or ruderal vegetation.

Some of the most typical anthrosols are the *plaggen* soils, which are created by mixtures of manure, organic waste, and mineral matter laid on sandy soils typical of the northern European lowlands (Holliday 2016). Other typical European anthrosols are the urban *dark earths* that are found in urban stratigraphic layers dating to the Middle Ages (see discussion in Case 12.1). Another anthrosol that has recently received a significant amount of research interest is Amazonian black earth, more typically known as *terra preta*, which was formed under conditions of continual slash-and-burn farming in the tropical rainforest areas of South America (Woods and Glaser 2004; Holliday 2016).

Anthrosols with relatively fast development include those created on raised fields in swampy and lacustrine areas. Typical cases of such soils include those the tropical lowland wetlands (*bajos*) of Mesoamerica (see discussion in Case 10.2) and the chinampas in the lacustrine and palustrine areas of the Mexican highlands (Case 11.2). The formation of such soils is based on the creation of beds in shallow waters and the constant accumulation of organic-rich sediment formed in the palustrine or lacustrine bed. As the organic sediment is excavated it forms canals that eventually divide the raised fields and are used for access to the fields by canoe.

Soils accumulated behind agricultural terraces and check dams are created by capturing soils transported by runoff or overland flow, or directly by bringing in soil from elsewhere (Donkin 1979). But in other cases, soils have to be carried out to make new cultivable areas on barren surfaces, as is the case of the highly erodible soils of the hardened volcanic tuffs on the slopes of central Mexico (Cordova and Parsons 1991; Borejsza and Frederick 2010).

Identification and anthropogenic features in soils

Chemical and physical properties of soils can bear testimony of past farming even if such soils have not been farmed for centuries, as it occurred in the American Southwest where indication of their existence was given away by former rock structures built to preserve moisture (Sandor and Eash 1991; Homburg and Sandor 2011)

The Crimean Plains still bear the imprint of ancient Greek farming more than two millennia ago (Lisettskii, Smekalova, and Marinina 2016). Many of these areas until recently were considered virgin steppe, because rainfall is insufficient for crops and irrigation until recently was not possible. But detailed archaeological survey supported the idea of occupation during a relatively wet climatic phase in the first millennium BC (see reference in Case 11.1, Chapter 11).

Typical features of agriculture such as plow horizon (Ap) may not make soils anthrosols, but do indicate human influence. Plowing of floodplain soils with intermittent overbank accumulation may transform the typical bedding of cumulic soils into a massive, still organic-rich horizon, that qualifies as a low horizon (Ap), a feature still distinctive in some paleosols. Agriculture intensification may lead to the formation of an *agric horizon*, a subsurface alluvial horizon, identifiable below the plow horizon (Bockenheim 2014). All these features can be preserved in paleosols, indicative of anthropogenic processes.

Soil micromorphology can help identify some anthropogenic influences in soil formation that are not visible to the eye. Anthropogenic indicators are less clear in situations where human influence on soil formation is very slight. Sometimes only certain proxies, as well as the history of the soil, may suggest so, as is the case of the prairie soils discussed in Case 12.2.

Soils as reflection of culture and cultural heritage

In view of the many human influences on soil formation and the importance of soil in cultural development, soils are not only natural objects, but also part of the cultural heritage (Blum, Warkentin, and Frossard 2006). Certain types of soils are not only the basis for a particular cultural group, but also that group has shaped the soil. As an example, take the case of raised fields created by Mesoamericcan cultures (e.g., the Maya raised fields in the bajos and the chinampas of the Basin of Mexico), or soils formed behind terraces as it occurs in the Andes.

Another argument that supports the idea of soils as cultural heritage is that archaeological and modern traditional agricultural soils as are part of the cultural landscape. This is not necessarily the idea of soilscape discussed above, but is the idea that soils are one of the features of archaeological landscapes. In this sense, the idea of protection is an important issue, for which geoarchaeology can provide recommendations.

CASE 12.1: PASTURELAND, CROPLAND, AND OTHER PAST HUMAN ACTIVITIES IN THE GEOARCHAEOLOGICAL RECORD BELOW BRUSSELS: A LOOK INTO URBAN DARK EARTHS

European dark earths: A background

The dark earth anthrosols found in some layers in the stratigraphy of some European cities have been studied considerably in recent decades. But despite the amount of research, there is still no consensus as to the causes of their formation, mainly because several causes converge in what is known collectively as *dark earths* – a typical case of equifinality. In view of this situation, it has been suggested that dark earths should be considered just a provisional term before it is replaced by a more descriptive one (MacPhail, Galinié, and Verhaeghe 2003). In the attempt to use a more scientific term, *urbic anthrosol* has been proposed (Blume 1989), but the problem is that in some cases these soils were not formed in urban environments and in others they are not even soils but cultural deposits (Nicosia and Devos 2014).

Despite their diversity, the European dark earths have a few traits in common. First, they lie predominantly below modern and recent urban layers and their ages range between the late Roman period and the late Middle Age (Nicosia and Devos 2014). Although most examples have been studied in Eastern Europe, they have been reported in southern Europe (Nicosia and Dvos 2014), central and Eastern Europe (Krupski et al. 2017), and European Russia (Alexandrovskaya and Alexandrovsky 2000; Dolgikh 2011).

In cities with Roman occupation, such as London and Paris, the traditional explanation to their origin has pointed to the ruralization of former urban spaces caused by the decline of Roman cities, although a number of studies show that dark earths are associated with continuous occupation (Mac Phail, Galinié, and Verhaeghe 2003; MacPhail, Nicosia, and Devos 2014). Furthermore, dark earths are found in cities that postdate the Roman period, or were never occupied by the Romans, such as those in Eastern Europe. This has led many researchers to interpret the formation of the dark earths as the result of a particular form of urban land use during the Middle Ages (Mac Phail, Galinié, and Verhaeghe 2003; Vrydaghs et al. 2016; Krupski et al. 2017). However, results from numerous studies show that in some cases the dark earths are not soils, but urban sediments (e.g., Krupski; Dolgikh 2011). Therefore, it has been proposed that their origin is a matter of case by case (Mac Phail, Galinié, and Verhaeghe 2003).

Examples of dark earths from two localities in central Brussels

A combination of archaeological, archaeopedological, soil microstratigraphy, and phytolith research provides a detailed insight into dark earth layers in two localities in downtown Brussels: Rue Dinant (Devos et al. 2009) and the garden in the Court of Hoogstraeten (Figure 12.3a). Both localities are within the thirteenth-century city wall, but with a sequence showing soil predating the construction of the wall.

The locality at Rue Dinant (Figure 12.3b) shows one level of dark soil that, rather than having one period of formation, has at least several phases, one of which starts as a grassland developed on a soil directly on an erosional surface formed before the agricultural settlement. At Locus 13, grassland soil has the characteristic biogenic porosity created by earth worms and other characteristics of meadow soils with considerable inputs of manure, suggesting its importance in a grazing landscape. The soils under the rampart (Zone 1–2) have inputs of sod soil, as to create a new soil on the eroded surfaces. Both soils are subsequently plowed, but eventually capped by layers that are more characteristic of hearths and urban occupation layers (Devos et al. 2009).

A similar grassland soil that formed under the same characteristics was identified in the Court of Hoogstraeten profiles, referred to as the lower dark earth (Figure 12.3c). This grassland soil, which served as pasture, was subsequently plowed, as shown in fragments of turf material, clay coatings, and phytolith remains of cereals (Figure 12.4a–b). This development coincides with a phase of agricultural expansion around Brussels between the tenth to thirteenth centuries (Devos et al. 2013; Vrydaghs et al. 2016). Subsequent change in land use is identified in microstratification of particles (Figure (12.4c–d), high phosphorous content, and the degree of articulation of grass phytoliths identify it as urban land use. An upper dark earth has been identified as a garden soil with domestic artisan waste and poor presence of grass phytoliths. This soil is associated with the gardens of the historically documented sixteenth-century Court of Hoogstraaten (Devos et al. 2013).

The stratigraphy of the sections that have been studied shows some parallels particularly in the origin of the soil, which apparently developed from an eroded surface where a grassland was established, perhaps by deposition of material eroded from the slopes (Devos et al. 2013). The agricultural phases of the dark earths in all profiles show remains of mortar and bone and high levels of phosphorous, suggesting that the areas were fertilized with

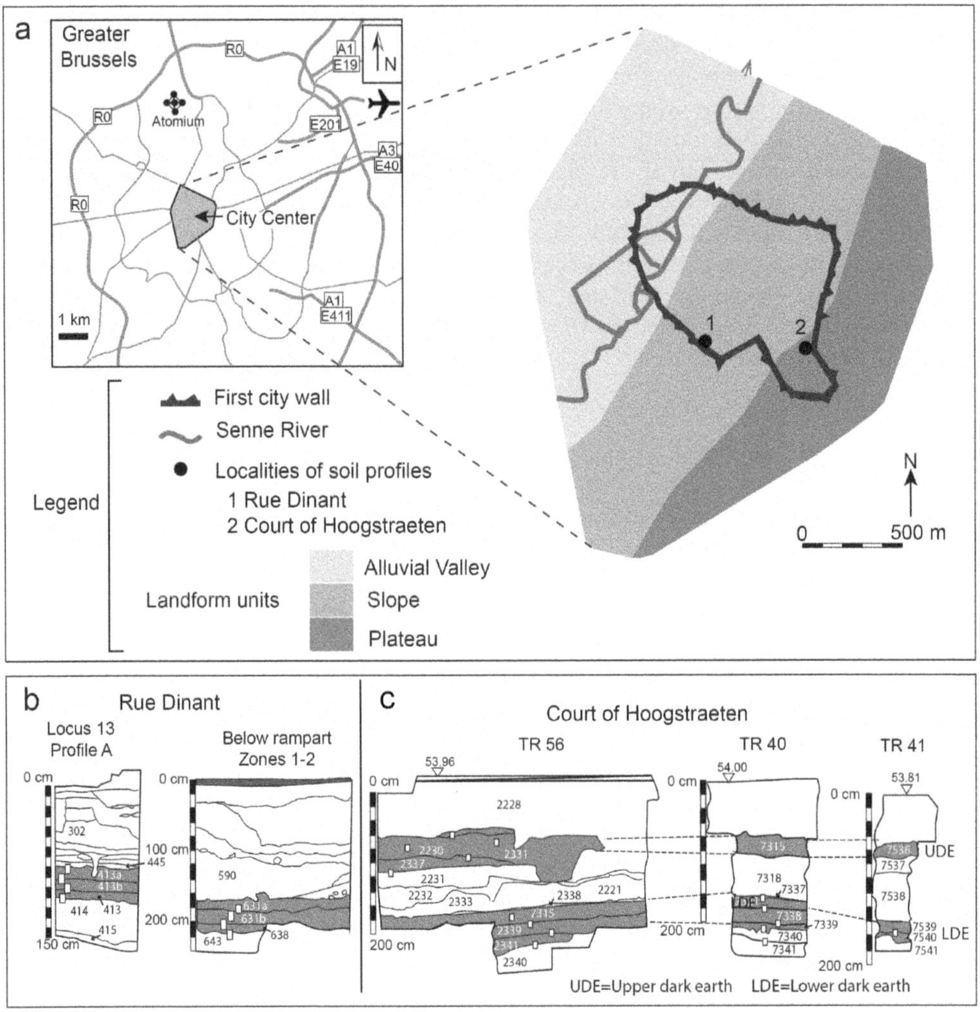

Figure 12.3. a) Location of soil profiles in the context of the thirteenth-century city walls, and stratigraphic profiles. b) Soils at Rue Dinant and c) the Court of Hoogstraeten. Modified from Devos et al. (2009, 2013).

domestic waste and manure and sometimes burned material. Another aspect that is worth mentioning is that at one location dark earths show a development from grassland/pasture to agricultural land to areas of urban garbage, and eventually to urban land use, which in turn supports the polygenetic nature of the dark earths (Devos et al. 2009).

A study of the dark earths in Brussels permits understanding not only of the evolution of the city, but also the medieval management of the land during the different phases of rural and suburban develoment. These phases constitute an important piece of information on medieval agricultural practices, whose counterparts in the countryside have been erased by plowing (Vrydaghs et al. 2016).

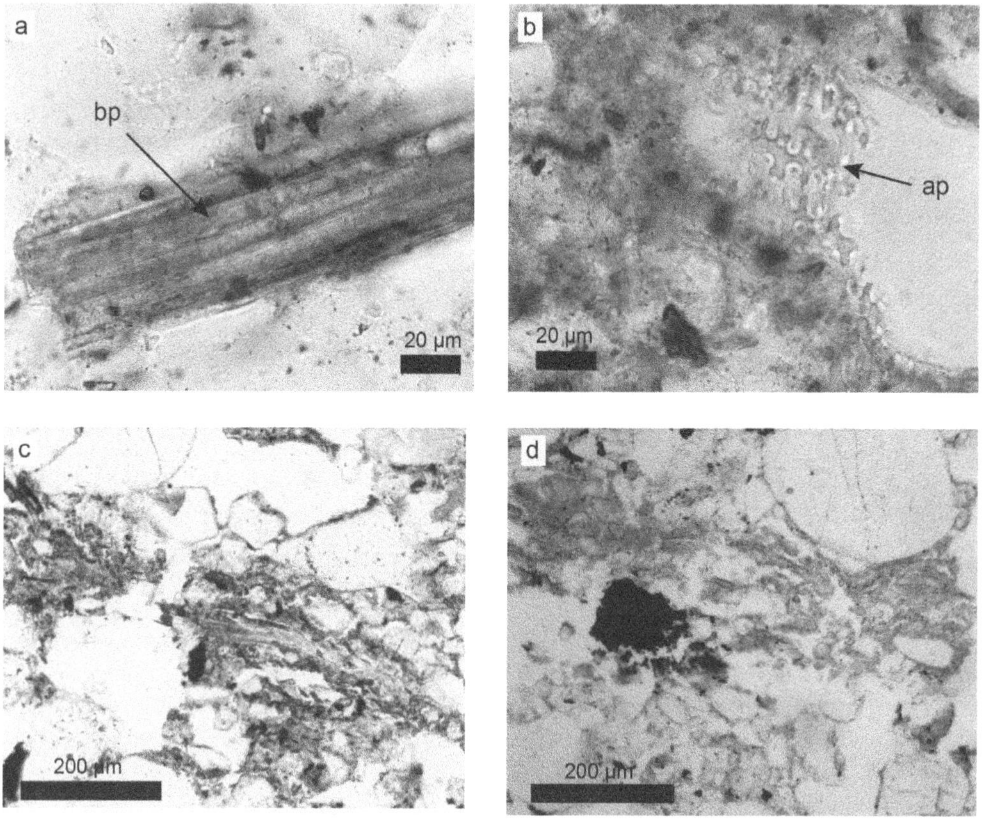

Figure 12.4. Selected micromorphology images of dark earths at the Court of Hoogstraeten: lower dark earth (layers 7321–7338), a) blackened grass phytoliths (bp); b) an articulated dendritic phytolith of cereal husk (ap), in dirty clay coatings with interstitial dust, and; c and d) potential occupation floor (layer 7337), characterized by micro-layering and fragments of domestic waste. Photos provided by Luc Vrydaghs and Yanick Devos from the Centre de Recherches d'Archéologie et Patrimoine, Université Libre de Bruxelles, Brussels, Belgium.

CASE 12.2: THE UNDERSTATED HUMAN INFLUENCE ON NORTH AMERICAN PRAIRIE SOILS: THE CONCEPT OF BISON PALEOPASTURES

Background

It is known to archaeologists, environmental historians, and ecologists in the Great Plains that native groups in prehistoric times used fire for management of the prairies, particularly for eliminating woody plants and promoting the growth of grass (Vale 2002). Purportedly, the renewal of grass cover through fire would attract large grazers (i.e., bison) to areas where they could be hunted (Stewart 2003; Vale 2002). If one considers that such practices go back several millennia, it is possible to think that constant burning and grazing by a dominant herbivore may have had an influence on soil development.

In the modern ecological and conservationist view, management of prairies with fire and bison grazing is seen as the *natural way*. Therefore, prescribed burning is applied to the prairies for maintaining biodiversity and habitats for certain prairie species. But as explained above, the practice is not completely *natural* as such prehistoric ecosystems have evolved over at least ten millennia of human-environmental interaction.

The Great Plains grasslands, locally known as prairies, occupy a vast area of the continent containing several grassland types, such as the shortgrass, tallgrass, mixed, and other combinations (see Figure 8.5). The tallgrass prairie is the westernmost type of grassland in the Great Plains, where it intermingles with forested ecosystems to form an ecotone characterized by mosaics of prairie and forests. From a long-term biogeographic point of view, it is understood that fluctuations of climate would permit advances and retreat of forests into what is now the tallgrass prairie (Anderson 2006; Cordova et al. 2011). This suggests an interesting hypothesis, first proposed for the part of the tallgrass prairie called the Prairie Peninsula, which states that this ecosystem was largely created by fire (see history in Anderson 2006).

The Prairie Peninsula covers most of the state of Illinois and parts of the states of Iowa, Indiana, and southern Wisconsin (Figure 8.5). Because it is an area of grassland that protrudes into the Eastern Deciduous Forests, it has been seen as an anomaly, suggesting that humans created it through constant burning. However, pollen and other paleoecological data point to the formation of the Prairie Peninsula as the combined action of climatic and human fires (Nelson et al. 2006). Nonetheless, no studies focusing on soils at large scale have looked at the traces of forest or grassland.

In terms of grass composition and relation with forests, the Prairie Peninsula shares many ecological characteristics with the tallgrass prairies, suggesting that the later may have also been formed under the same long-term processes. Unlike the Prairie Peninsula, the tallgrass prairie in the south and central Great Plains states does not have lakes that could add information on past developments. Thus, one particular way to test some hypotheses concerning the evolution of the prairie is through proxies from soils (Cordova and Johnson 2009; Cordova et al. 2011). But this actually brings another question: Was the tallgrass prairie the same through the Holocene?

Hypotheses and implications

The question whether the tallgrass prairie existed through the Holocene or if it is a more recent phenomenon is one aspect that appears from a study of plant microfossils and carbon isotopes from soils in the south-central Great Plains (Cordova et al. 2011). Data from this study shows that while the short and mixed-grass prairie have existed since the early Holocene, records of the tallgrass prairie are fuzzy. Only in the past 2000–3000 years, assemblages of phytoliths are comparable to those of the historic and modern tallgrass prairie (Cordova et al. 2011). Records of bison in the early and middle Holocene, despite the gaps, are more common in what are now the shortgrass and mixed-grass prairies, but not in the tallgrass prairie (author's current research).

These sets of data have prompted the idea that perhaps the tallgrass prairie is the result of climatic and human changes that occurred after the Hypsithermal, that is to say, towards the end of the Archaic period of the Plains (see Case 8.2). It is possible that changes in cultural patterns probably related to a mixed economy combining horticulture and bison hunting caused that change. This hypothesis considered that increase in effective moisture after the Hypsithermal may have also been important in using the tallgrass prairie as a landscape for both hunting and farming. Thus, attracting bison from their optimal habitat (the short-grass and mixed-grass prairies) to the more marginal tallgrass prairie would have required transforming the landscape into a more attractive grassy environment.

A methodological model

Testing the hypothesis concerning the development of the tallgrass prairie under the influence of human-led frequent fires and bison grazing requires consideration of many ecological changes that can be deduced from the archaeological and historical record. However, on the parts of the geoarchaeological and paleoecological records obtainable from soils, it is possible to study such an influence using proxies obtained from soil and paleosol sequences across a region, a methodological scheme named *Paleobiomes, Paleopastures, and Paleofires* (Cordova and Johnson 2007; Cordova et al. 2011). The idea is to use proxies for vegetation structure, megaherbivore density, and fire incidence obtained from soil horizons, whether top-soil or buried as paleosols (e.g., Figure 12.5a).

Using carbon isotopes and phytoliths from soils it is possible to reconstruct vegetation structure, complemented by other proxies, particularly those of general fires (charred particle concentration), grass fires (burnt grass phytoliths), and the relative density of herbivores (concentration of coprophile spores) (Figure 12.5). Opal phytoliths and $\delta^{13}C$ from the soil organic matter can help also reconstruct the structure of vegetation, particularly the relative dominance of grasses and woods, as well as assess the dominance of cool season (C_3) versus warm season (C_4) grasses, and among the latter those that are mesic (Panicoideae) from those that are drought-resistant (Chloridoideae) (Cordova et al. 2011). Additionally, on-grass phytoliths, particularly those associated with wooded plants, are useful as means of reconstructing the incidence of trees and shrubs.

Some of the localities in the preserves managed with fires also had sections with soils that could be dated, as in the examples of Figures 12.6, 12. 7, and 12.8. The proposed study is meant to encompass modern data from different areas managed with prescribed fires and bison, where the frequency of fire and bison density are known. As shown in the pilot study, data from bison, cattle, and no-large-herbivore areas can also be useful in determining the influence of large herbivores (see details in Cordova et al. 2011).

Prospects and applications

The cross-correlation of proxy data obtained from soils on soils can be placed in a spacetime model using GIS (Figure 12.5b). Categories of soils can be divided depending on location

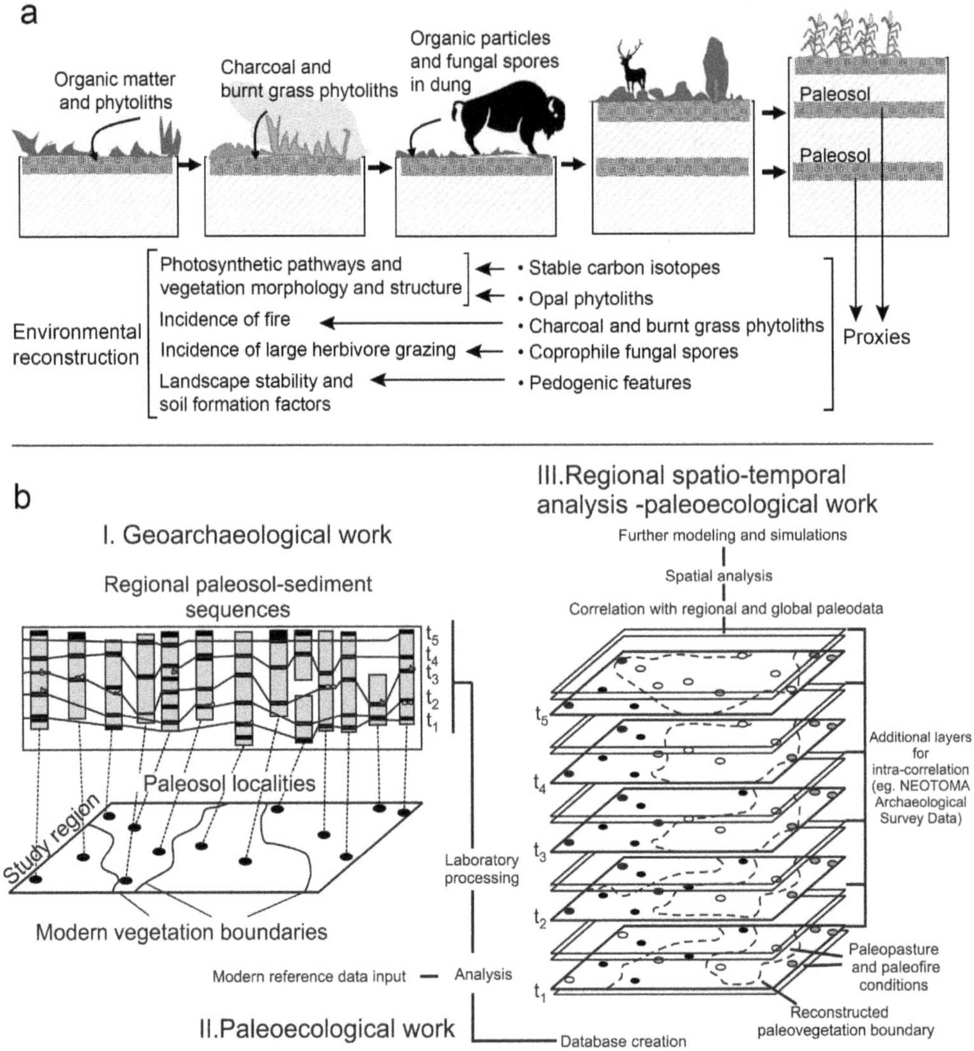

Figure 12.5. (a) Diagrammatic model of evolution of the Great Plains prairie soils and paleosols under the influence of herbivores and fire and proxies that can be obtained from paleosols, and (b) model for the collection and analysis of proxy data at the level of site and setting and its further spatio-temporal analysis at contextual levels of landscape and environment.

Figure 12.6. Section and setting of soil profile studied at the Tallgrass Prairie Preserve, Osage County, Oklahoma.

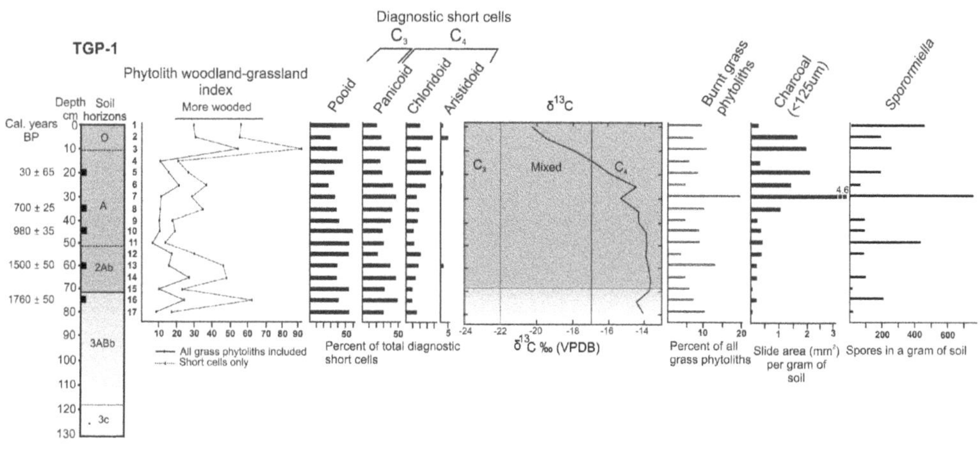

Figure 12.7. Example of soil proxy dataset for the same soil profile in Figure 12.6 with data. Modified from Cordova et al. (2011).

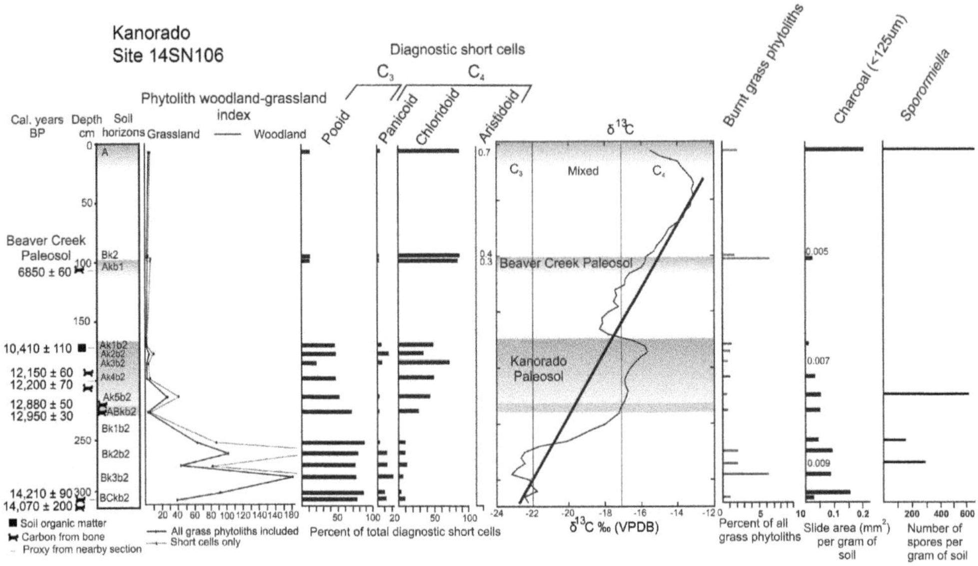

Figure 12.8. Example of paleopasture datasets in a soil in the Kanorado Site, Kansas. Modified from Cordova et al. (2011).

(flood plain, terraces, uplands, loess sequences, etc.). The spatio-temporal model can provide a diachronic scene to bison paleoecology that could help explain the landscape and environmental context of the cultures that inhabited the region.

Such a project should also provide information for establishing baselines for conservation in the grassland reserves and mixed forest-grassland reserves in the region. This suggests potential collaborations with researchers, policy-makers, and the public concerned with the conservation of natural and cultural heritage landscapes.

13

The Geoarchaeology of Natural Disasters

NATURAL DISASTERS IN THE HUMAN-ENVIRONMENTAL CONTEXT

The term *natural disaster* is often flexible as to designating destructive phenomena occurring in a short time span (e.g., earthquake, tsunami, and tornado) and phenomena of more protracted duration (e.g., droughts, inundation of territories by sinking ground or rising sea levels, drying of a lake, or even the successive occurrence of calamitous weather phenomena). The term *natural catastrophe* is used with the same meaning, i.e., synonymous with natural disaster.

In geology, a catastrophe is often interpreted as an event that causes spectacular effects, such as volcanic eruptions, tsunamis, and meteorite impacts (Engelhardt and Zimmermann 1982). In geomorphology, a catastrophe is often an extreme or high-magnitude event process occurring in a short time (Brunsden 1996). But the spectacular effects, the magnitude, and the duration of the event are a matter of scale and perspective. From the human point of view catastrophes are more often defined by the impact they have on society and its infrastructure.

Apart from the perspective and scale factors involved in the definition of a catastrophe or disasters, the matter of causality is also important. It is understood that natural disasters or catastrophes are caused by *natural* phenomena such as volcanic eruptions, earthquakes, tsunamis, tornados, and hurricanes. They differ from human or technological disasters, which are caused by negligence, mismanagement, or any other human cause (e.g., war). However, natural disasters can trigger a series of other collateral events that may not necessarily be natural phenomena, but may be combined with human-led causes. Sometimes such a combination of natural and human events may constitute an *environmental crisis* (see Chapter 14).

The term environmental disaster is often found in the literature. They differ from natural disasters in that they are caused by an intricate combination of natural and social phenomena. Therefore, they are essentially environmental crises. Their designation as disasters is a matter of perception, particularly when they produce high-magnitude effects in a short time. For example, the Dust Bowl of the 1930s is often referred to as an environmental disaster, even though it was an environmental crisis that lasted longer than just the decade of the 1930s (see Case 14.2).

Cases of geoarchaeological studies of past natural disasters abound, and the approaches, methods, and techniques to study them vary considerably depending on the case and the nature of the evidence (i.e., the record). The two cases selected for this chapter deal with the approaches to and interpretation of the prehistoric Xitle eruption in the Basin of Mexico (Case 13.1) and the formation processes and sediments of flooding in New Orleans during the Hurricane Katrina disaster in 2005 (Case 13.2). The first one illustrates some of the problems in reconstructing causality in time when the record is fragmentary. The second is an actualistic approach to interpreting natural and cultural processes of sedimentation in an urban environment.

GEOARCHAEOLOGICAL APPROACHES TO NATURAL DISASTERS

General issues in natural disaster research

Natural disasters in the past have drawn the attention of archaeologists because of their influence on ancient societal dynamics, including the explanation of the collapse of some civilizations. In this context, geoarchaeology provides the tools to reconstruct the timing and extent of events involved in the development of a disaster.

Unlike the core geoscience methodological approaches and techniques to study natural disasters, geoarchaeology looks at natural disasters within a human framework. This means also the understanding of aspects of culture, the archaeological record, and in general the societal context of the disaster. Therefore, at the same time that geoarchaeology provides tools to answer questions of a geological and geomorphological nature, it also investigates aspects of human-environmental interactions that characterize the relations of people to disasters, as is the case of vulnerability, risk, response, and resilience, among others.

Natural disasters vary in speed, predictability, duration, and frequency, aspects which are important for understanding the human processes of human adaptation and resilience to human disasters (Johnson 2017). Therefore, the application of geoarchaeological methods and techniques depends on the type of disaster, the occurrence in time and space.

Another significant factor in the geoarchaeological study of past natural disasters is the amount of information available for the reconstruction of events. Generally speaking, historical disasters are easier to study because of the amount of information available. Because the historical period in most parts of the world is recent, evidence of a catastrophic event and its effects on human populations is more likely to be preserved. But there are also numerous cases in late prehistoric times where evidence of disasters and their effects on societies is preserved.

Natural disasters and their effects on human populations in deep prehistory are more difficult to reconstruct because of the fragmentary nature of the record. This situation leads to uncertainties and reconstruction of facts at the level of assumption, for example, in the effects of the Mount Toba megaeruption (in the Indonesian Archipelago) around 75,000 years BP. This eruption has received attention because of its putative implication in changing the environment and hominin populations at a time when anatomically modern humans began to disperse out across southern Asia (Ambrose 1998). The magnitude of the event, estimated to have no analog in historical time, is believed to have impacted climate and ecosystems for several centuries, leading to the formation of a genetic bottleneck in humans. However, research results from the numerous studies of the event and the archaeological record are largely inconclusive and not able to provide strong evidence to the presumed effects on humans (Williams 2012).

Another case of a prehistoric volcanic disaster is that of the eruption of a volcano in southern Italy that produced the Campanian Ignimbrite around 40,000 years ago. This eruption is thought to have deteriorated the climatic conditions and ecosystems of Europe, thus impacting the populations of Neanderthals and delaying the migrations of modern humans into Europe (Fitzsimmons et al. 2013). The reconstruction of the eruption and its effects have been thoroughly studied using *cryptotephras* in different settings, including archaeological sites, and through the implementation of models using the stratigraphic and archaeological data (Fitzsimmons et al. 2013; Marti et al. 2016). The smaller distribution of the effects, the better-preserved record, and the more coordinated research have produced more data for this case than for that of the Mount Toba eruption.

Unlike volcanic eruptions, other disasters are more difficult to reconstruct because they leave fewer traces in the record. These include, for example, extreme weather phenomena such as storms and tsunamis, and earthquakes, among others. The evidence of past storms and tsunamis, as discussed below, are now recently being studied, but only very recent events been identified in the record. The study of earthquakes in the past is much easier when there are human structures to date, which is the aspect that defines the difference between archaeoseismology and paleoseismology (Niemi 2016).

The fragmentation of the record is often the primary problem in reconstructing natural catastrophes and their impact on humans. The invisible and absent parts of the record, however, can be accounted for, either through the study of traces (e.g., cryptotephras), methods to study indirect evidence (e.g., records in lake sediments), and modeling. Techniques are available, but what is important is the interpretation of the data, which is a topic that is the central subject of this chapter.

Interpretive approaches

The interpretive approaches to reconstructing past natural disasters and their effects on human populations depend more on the strategies for collecting, assembling, correlating data, and reconstructing scenarios. In most geoarchaeological research of prehistoric disasters, particularly those for which the record is fragmented, the establishment of hypothetical scenarios are built upon existing data, sometimes with the help of processes observed in modern reference events, i.e., the actualistic approach. Such events may include volcanic eruptions, floods, earthquakes, tsunamis, and storms (see the case of paleotempestology below). Additionally, computer modeling can help reconstruct scenarios through exercises of simulation.

In general, geomorphological studies of recent catastrophic phenomena provide important insights into the formation of deposits, insights which can be used as a reference for studies of similar deposits of the past. In the USA, studies of sediments and landscapes of recent catastrophes that have left a record for the future have been studied and reported, as is the case of the lahars of the Mount Saint Helens eruption of 1980 (McEwen and Malin 1989) and the overbank deposits of the 1993 floods in the upper Mississippi (Gomez et al. 1997), among others.

An example of an actualistic study in New Orleans (Case 13.1), which was not originally done with a geoarchaeological purpose, provides an insight into the formation of the record under flood conditions in large rivers. Similarly, the Dust Bowl formation processes in the eolian record of the Great Plains is another case to cite (Case 14.2). Perhaps these examples will encourage similar studies in places where disasters have recently taken place.

Paleotempestology: An emerging approach with applications in geoarchaeology

Paleotempestology is a relatively young approach that is defined as the study of the evidence of past tropical cyclones in sedimentary records in coastal depositional environments (Liu and Fearn 2000). The field was developed at the crossroads of sedimentology and coastal geomorphology, and to a lesser extent of climatology, meteorology, and environmental history. Over the years, however, the field has added to its strategies other methods such as the study of coral-rings, speleothems, tree-rings, and historical documents, among other sources. Also, its scope has gone beyond the tropics to study the records of storms at other latitudes.

The sedimentary evidence of past storms (and particular tropical cyclones) is formed in lagoonal environments, where sudden depositional events of sand (overwash) are associated with the occurrence of single-event violent storms. These sand layers are often interbedded with typical lagoonal, clay and organic, sediment. The sedimentological approach is also combined with the presence and abundance of foraminifera in sediments in water bodies and swamps, as evidence that sea surges created by hurricanes penetrated areas normally fed by terrestrial waters (Liu and Fearn 2000; Wallace and Anderson 2000).

Although amply criticized from its beginnings due to possible misrepresentations of the stratigraphy created by other phenomena (e.g., tsunamis), and failures in detecting low category hurricanes and tropical storms, paleotempestology has established itself through persistent solid research (Otvos 2011; Muller et al. 2017). The field is evolving rapidly as more technological approaches and evidence are generated in different parts of the world.

Although paleotempestology was born outside archaeology and geoarchaeological research, paleo-storm records from the Gulf of Mexico and the Caribbean are now being cited as background information to correlate with historical and archaeological events (Cooper 2012). Some studies have aimed at obtaining records to reconstruct prehistoric patterns of hurricane intensity that can be correlated with general circulation patterns and phases of intensification of global phenomena such as ENSO phases (Fan and Liu 2008; Wallace and Anderson 2010; Williams 2013). Sediment records of the coastal lagoons of Belize are now providing information that correlates with developments in the Maya region (McCloskey and Keller 2009; Beach et al. 2009). In this sense, paleotempestology is an important potential tool in geoarchaeology particularly, at the contextual level of environment.

CONTEXTUAL LEVELS AND ISSUES OF INTERPRETATION

The effects of natural disasters can be studied at all contextual levels, but each level represents a different methodological and interpretive approach, an aspect that must be taken into account to avoid generalizations. Sometimes evidence of a disaster is better discerned at the level of environment, particularly because certain records, such as lake deposits, tend to better preserve the evidence of a phenomenon. In other cases, disasters are more discernible at the level of site, as is the case of ancient earthquakes, where damage to structures, collapsed structures, and associated artifacts can help date the event (Niemi 2016).

One interesting recurrence – although it can be considered a rule – is that the best contexts for studying natural disasters are those that were abandoned after the disaster, or even better those that were sealed (e.g., Pompeii and Herculaneum in Italy and Joya del Ceren in El Salvador). In contrast, in those sites that have been re-occupied or rebuilt, evidence and traces of the catastrophe have been removed.

Joya del Ceren is an excellent example of a well-preserved moment of eruption, a reason the site is dubbed the Latin American Pompeii (Figure 13.1). It is a unique case in the region in the sense that preservation is such that behavioral aspects before and as a consequence of the eruption can be reconstructed (Sheets 2006). Inside the houses, food was found in pots, some of which were protected as the roofs collapsed (Figure 13.1a), and in some cases entire fields cultivated with maize and other crops were recovered (Figure 13.1b).

But not all cases where sites are sealed are easy to study, as is the case of the record below the Xitle lava (Case 13.1). In some cases, sealed sites might represent a problem of visibility to the point that they may never be discovered. Not only ashes can bury a site; very often also landslides, mudflows, and massive flood deposits can contribute to the sealing of an occupation.

The best strategy for studying catastrophes in the past is perhaps to combine information from sites, settings, and landscape in the context of a solid environmental background. As mentioned above, the case of the impact of the Campanian Ignimbrite on modern human migrations and Neanderthals in Europe has been successful in part because of technologies applied to it, and in part because the study focuses on different contextual levels from archaeological sites to the environment (see Fitzsimmons et al. 2013; Marti et al. 2016).

CASE 13.1: THE XITLE VOLCANO CATASTROPHE AND ITS IMPACT ON THE PRECLASSIC AND CLASSIC ENVIRONMENTAL CONTEXTS IN THE BASIN OF MEXICO

Archaeological background

The site of Cuicuilco, located in the southern part of the Basin of Mexico (Figure 11.5), is an important landmark in the cultural evolution of Mesoamerica because it is one of its earliest urban settlements (Sanders, Parsons, and Santley 1979), despite the fact that its real size and distribution is not known because most of it is believed to be under the lavas of Xitli Volcano. Its foundation occurred most likely during the Tetelpan phase (800–700 BC) during the early Middle Formative (Preclassic) period (Carballo 2017). Estimates of its population vary among scholars, ranging between 10,000 and 20,000 during its apogee, probably in the Terminal Formative (Carballo 2017). The site was abandoned at the end of the Terminal Formative, or during the Early Classic, depending on most accepted dates for the eruption of Xitle Volcano, which is one of the main debates regarding the timing of natural and social phenomena related to this particular event.

The cone and main vent of Xitle Volcano is located eight kilometers south of Cuicuilco, and its lava extends over a large area now part of the southern suburbs of Mexico City

Figure 13.1. Joya del Ceren site, a) view of a preserved house with a collapsed roof, with a sequence of pyroclast layers in the background; b) farmed field rows. In some cases, remains of maize plants were preserved. Photographs by the author.

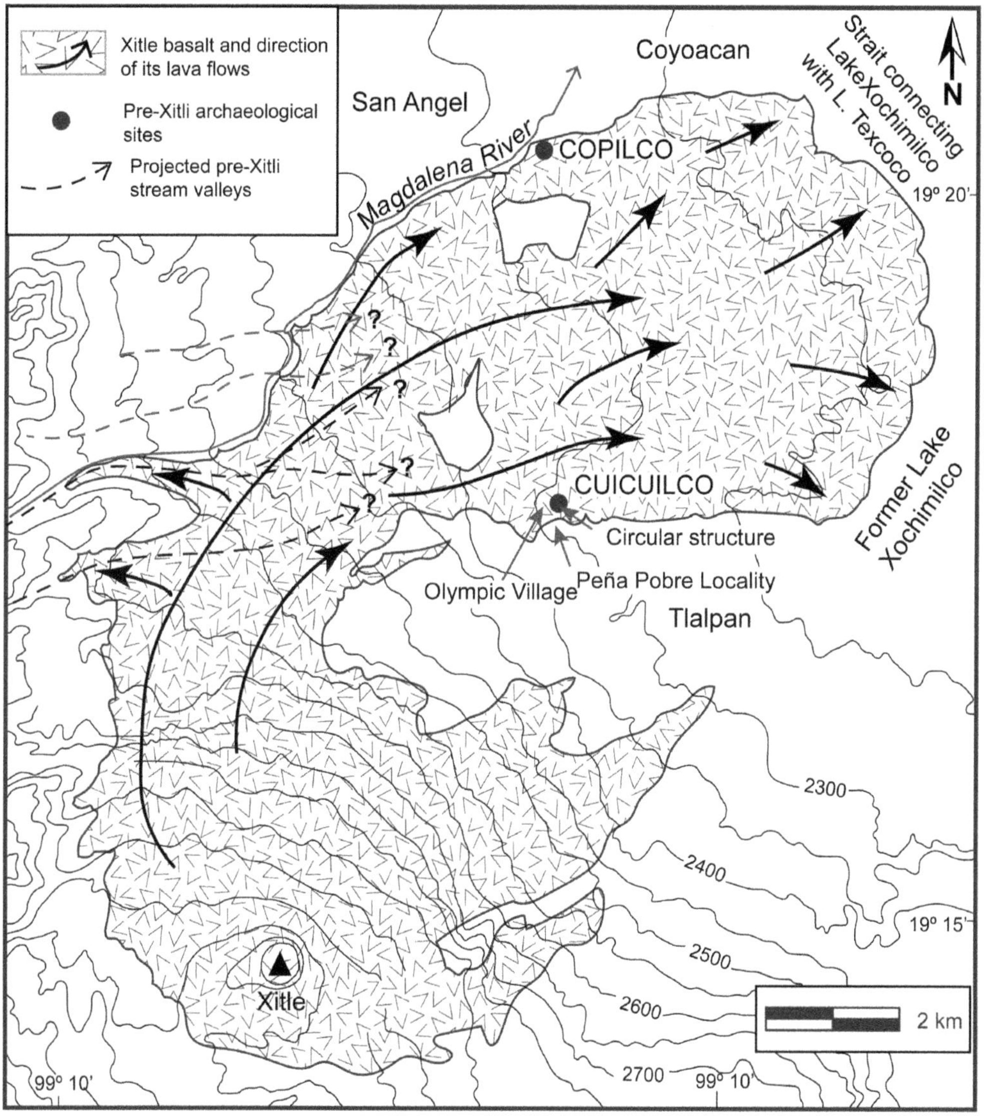

Figure 13.2. The Pedregal lava flow and localities mentioned in the text. Composed with information from Cordova, Martin del Pozzo and López-Camacho (1994); Delgado-Granados et al. (1998); Gonzalez et al. (2000); and Siebe (2000).

(Figure 13.2). Variations of the name Xitle are *Xitli*, or its original Nahuatl name *Xictli* (navel). The volcano is part of a field of cinder cones and basaltic lavas of Late Pleistocene and Holocene age (Martin del Pozzo, Cordova, and López-Camacho 1997; Siebe 2000).

The controversial aspects of the implications of the Xitli Eruption in the abandonment of Cuicuilco and the eventual rise of Teotihuacan have been discussed in Mesoamerican archaeology since the 1920s when Byron Cummings (Cummings 1926) exposed the main circular structure (Figure 13.3) from below the basaltic lava field known as El Pedregal.

Figure 13.3. Cuicuilco circular structure. Photo by the author.

Artifacts discovered in the circular structure seemed to correlate with another site discovered below the lava at the locality of Copilco, on the northern edge of the lava field (Figure 13.2), suggesting that under the lava were concealed remains of an ancient civilization. Subsequently, in the immediate vicinity of the main circular structure, at the Olympic Village (i.e., Cuicuilco-B), and at the Peña Pobre area, other structures were uncovered adding to the complexity of the so-called Cuicuilco urban civilization (Schavelzón 1982).

The question that remains about Cuicuilco is: when was the city abandoned? But because the default reason for abandonment was the eruption, then the first question to answer is: when did the eruption occur? The answers to these questions turn the attention to geology, and particularly volcanology, with a geoarchaeological approach.

A history of geoarchaeological research around Cuicuilco

The first attempt to date the eruption with the newly developed radiocarbon method provided a date of 2422 + /-250 ^{14}C years BP (Libby 1951). Subsequent research of monuments around the main circular structure provided more evidence for complexity of the event and its consequences in time and space (see the history of the research in Gonzalez et al. 2000).

In the 1990s, the study of Cuicuilco and the volcanic eruption was resuscitated through the study of structures and paleosurfaces exposed by the removal of the lava as the southern outskirts of Mexico City began an unprecedented urban development of the El Pedregal lava field (Cordova, Martin del Pozzo, and Lopez-Camacho 1994; Martin del Pozzo, Cordova and Lopez-Camacho; Pastrana 1997). These studies began to question some of the assumed ideas about the age of the eruption and the abandonment of Cuicuilco, mainly because of the variety of dates from surfaces below the lava.

In subsequent years, studies focused on obtaining more dates and obtaining more paleo-landscape information from areas where the basaltic cover was being removed, suggesting different times and scenarios for the eruption and its consequences on ancient populations in the region (Delagado-Granados et al. 1998; Gonzaelez et al. 2000; Siebe 2000; Lugo et al. 2001).

In more recent years, using archaeomagnetic methods, radiocarbon ages have been revised, producing date frames more consistent with the original dating of the eruption centered around the beginning of the first millennium CE (Urrutia-Fucugauchi et al. 2016), then conforming to the original date by Arnold and Libby (1951). But still no consensus exists regarding the timing of events, in part because of the complexity of some stratigraphic contexts below the lava.

Some deposits had been interpreted as degradation of the circular structure due to its abandonment years before the eruption (Cordova et al. 1994), an interpretation which was backed up by the relatively late dates (Figure 13.4), but which disagrees with the interpretation of cultural layers at the circular structure (Pastrana 1997). The idea of more than one eruption has been mentioned, but some studies that have focused on the contact between the lava and the surface have pointed to a relatively fast, single eruptive event characterized by various effusive phases (Delgado-Granados et al. 1998; Gonzalez et al. 2000; Siebe 2000; Lugo et al. 2001).

The disparity in the radiocarbon assays from layers and soils under the lava flow comprises a broad period encompassing about a millennium (Figure 13.4). The problem with dating the time of the eruption is that in many places the surface in contact with the lava has been scorched. In other cases, where the soil was protected by ash-fall deposits, the long-residence effect of organic matter in soils has produced older dates. However, beyond the disparity of dates there are aspects of a cultural and natural character at a more regional level that are also worth discussing.

Archaeological implications of recent geoarchaeological research

Despite the disparity, most dates from below the lava are no older than 400 CE, which has been considered the minimum age of the eruption (Siebe 2000). Nonetheless, that age is not consistent with the youngest ceramics recovered from the excavations. Another approach that has been suggested is averaging the dates (Gonzalez et al. 2000) (Figure 13.4). Averaging puts the time of the eruption around 200 CE, which is consistent not only with the latest ceramics below the lava, but also with the abandonment of other sites in the southern part of the Basin and the concomitant growth of Teotihuacan (Plunket and Uruñuela 2006). It has been suggested that around the time of the XItle eruption, a plinian-type eruption of the Popocatepetl Volcano in the southeastern part of the Basin (see location in Figure 11.5) may have already contributed to the exodus towards the north of the Basin (Siebe 2000; Plunket and Uruñuela 2006).

The eruption of Xitle, and most likely also the eruptions of Popocatepetl, may have had a broad impact on the terrestrial and lacustrine ecosystems of the southern Basin. In fact, many areas in the southern part of the Basin were not populated again in large numbers until the Postclassic period (Sanders, Parsons, and Santley 1979), which suggests that the resilience of the landscape after the eruption, or eruptions, took some time.

Reference analogs of the Xitle eruption

The mechanisms of the Xitle eruption are understood in part by the stratigraphy of the ashes and lavas, which show several eruptive phases in a relatively short time

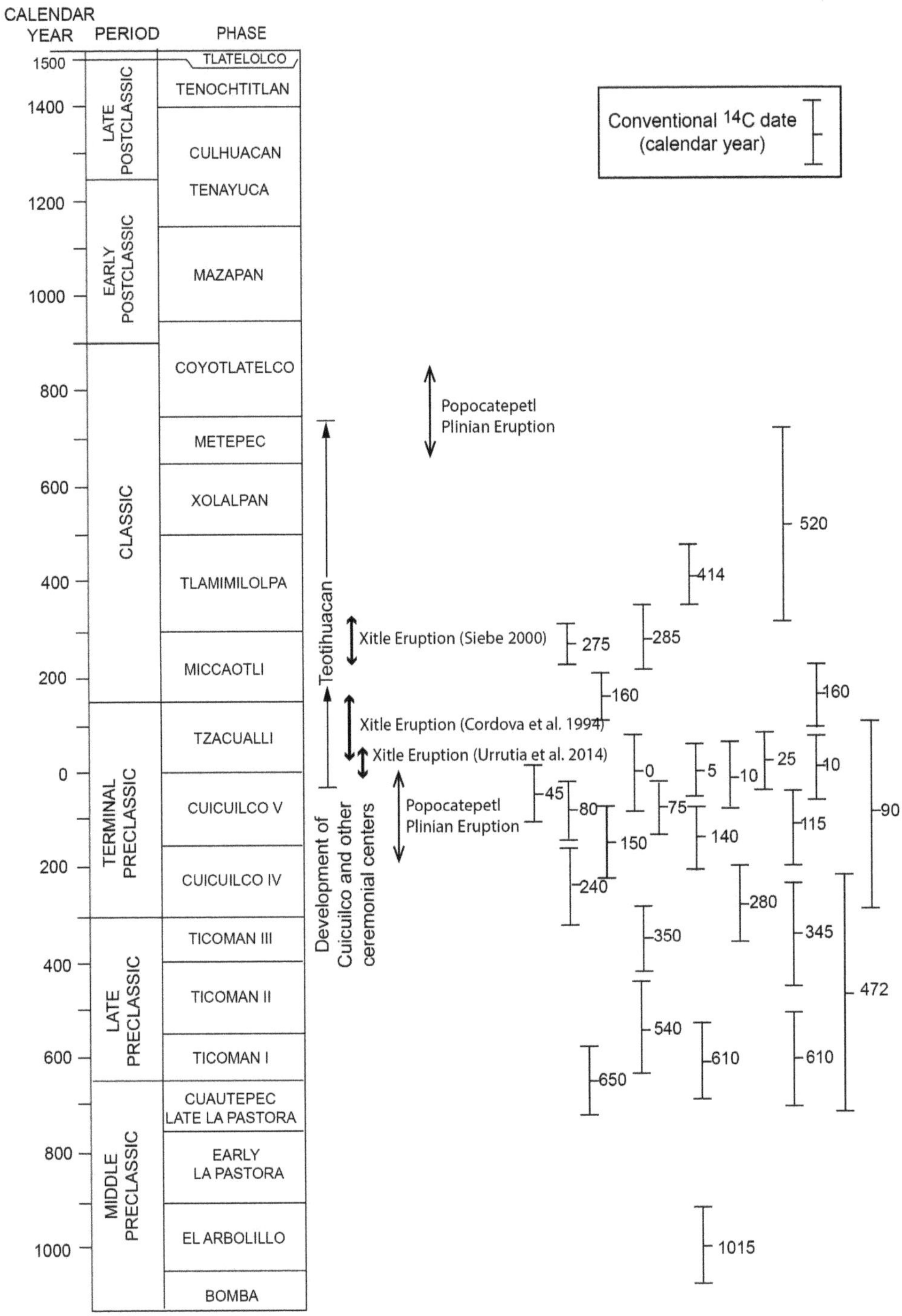

Figure 13.4. Archaeological chronology and radiocarbon dates (modified and updated from Gonzalez et al. 2000 and Siebe 2000).

(Delgado-Granados et al. 1998). A modern reference analog to this kind of eruption is that of Paricutin volcano, a cinder in the western Mexican state of Michoacán, whose eruption in 1943 encroached on several towns and forced populations out of the immediate area of the volcano. In tandem with ash fall and earthquakes, the lavas moved slowly enough so that the people could leave with their belongings, with no casualties (Nolan 1979; Rees 1979). But they could not resettle in areas covered by the lava, or even in areas covered by ash, because of the impossibility or difficulty of farming in these areas. Instead, they were resettled by the government in areas away from the damaged environments.

The eruption of another cinder cone, Sunset Crater (roughly 1050–1125 CE) in Arizona, provides an example of the impact of effusive (lava) volcanism on prehistoric populations (Pilles, Jr. 1979; Elson et al. 2007). The patterns of abandonment and lack of reoccupation also suggest a parallel with the Xitle volcanic event. But unlike the case of Paricutin, the ashes of the Sunset Crater volcano became more attractive to farmers as they preserved moisture in soils in an area of low precipitation (Pilles, Jr. 1979).

The eruption of Paricutin lasted about ten years, during which several lava flows alternated with ashfall (Rees 1979). The Sunset Crater eruption, despite disagreements with dendrochronology and dates, lasted several years, if not decades, amounting to perhaps no more than 50 years (Elson et al. 2007). Most likely a similar scenario occurred with the Xitle eruption, which may have lasted several years, thus complicating the transformation processes that today make it difficult to properly date and pinpoint the age of events.

One aspect that differs between Paricutin and Xitle is that ashes are not preserved in areas not covered by the lava. Thus, either the ash fall was less prominent in the Xitle eruption or covered a smaller area. There is also the possibility of its being washed away by runoff between eruptive events. If ash fell in smaller amounts, it was incorporated into the soils. But all these details are still to be studied, perhaps as part of the reconstruction of landscapes and environments before and after the eruption.

The paleolandscape of Cuicuilco

In their study of the sites affected by the Xitle lava, Cordova and colleagues (1994) attempted to reconstruct the pre-eruptive landscape by connecting landforms with those protruding from the lava and surfaces and sediments discovered below the lava. Likewise, determining the thickness of the lava also permitted reconstructing the paleotopography. The conclusion of the study, which was very sparse in terms of localities, pointed to several streams coming into the area and probably forming a delta in the western part of Lake Xochimilco (Figures 11.5 and 13.2).

Later studies added more information, particularly in relation to the existence of water, probably a lakeshore (Pastrana 1997; Gonzalez et al. 2000; Siebe 2000; Lugo et al. 2001) and soils in different locations (Solleiro-Rebolledo et al. 2016).

Although the pieces of information are minimal compared to areas not recorded below the lava, they indicate that Cuicuilco and its satellite settlements were located on a lakeshore, where several rivers provided water to maintain a high level of fresh water. The eruption blocked several rivers, diverting them to the north (to Lake Mexico or Lake Texcoco), probably reducing the amount of water in Lake Xochimilco (Cordova, Martin del Pozzo, and Lopez-Camacho 1994). Nonetheless, at this point the information on paleolandscapes and the mechanisms of the eruption are minimal, requiring the recovery of additional information. The difficulty is that a strategy has to be used because most of the areas with potential information are under the lava or under urban areas.

CASE 13.2: THE GEOARCHAEOLOGICAL RECORD OF HURRICANE KATRINA'S DISASTER IN NEW ORLEANS

Background

Hurricane Katrina developed in the Atlantic Ocean and, after passing over the Bahamas as a tropical storm and developing as a hurricane during its pass over the southern tip of the Florida peninsula, became a Category 5 hurricane (the highest category) before landing on the eastern Mississippi delta (Figure 13.5a) on the morning of August 29, 2005. By the time of landing, however, the hurricane had fallen to a Category 3, but it was still strong enough to cause damaging winds and a storm surge (28 ft. or 8.5 m.) that devastated a considerable area of the Mississippi delta and the northern coast of the Gulf of Mexico.

The storm surge penetrated into the delta, raising the level of the Mississippi river considerably and putting extreme pressure on the levees protecting the city of New Orleans. At the same time, the strong winds from the north caused massive waves on Lake Pontchartrain, north of the city, which also put pressure on the levees. As a result of such

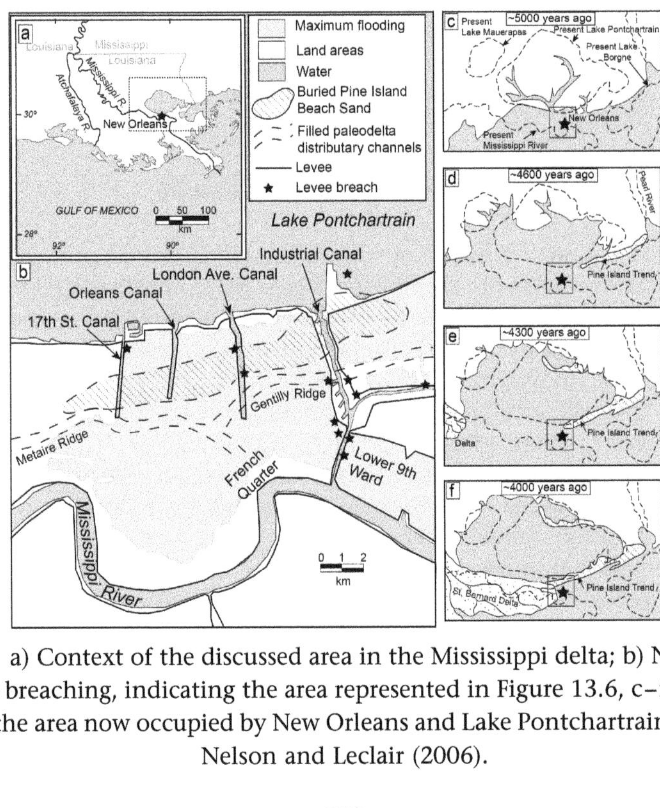

Figure 13.5. a) Context of the discussed area in the Mississippi delta; b) New Orleans and areas of levee breaching, indicating the area represented in Figure 13.6, c–f) Late Holocene evolution of the area now occupied by New Orleans and Lake Pontchartrain. Modified from Nelson and Leclair (2006).

pressures, the levees surrounding the city were breached at several points in the city (Figure 13.5b), causing considerable flooding with water and sediment into the streets and houses.

The flooding of the city caused widespread devastation, thousands of deaths and displaced people, and millions of dollars in losses. The city's population and economy declined in the years following the disaster, and has only recently begun to recover after costly upgrades of its levees and drainage system. The damage caused to the city was not solely caused by the natural disaster, i.e., the hurricane, but a series of events that the event unleashed. However, many of the causes of the destruction lie not only on the natural disaster, but in a series of human-led events that preceded the disaster.

New Orleans was originally founded on the natural levees of the Mississippi at the location of the *Vieux Carré* (Figure 13.5a), but as the city grew, it expanded beyond the levees onto backswamps. To prevent flooding in the city, the levees were upgraded and water was pumped out of the underground. But these measures were of no use under the conditions created by Hurricane Katrina. The problem also lay in a series of changes in the Mississippi river channel, particularly a dam upstream that was meant to prevent water from going into the Atchafalaya Channel, thus increasing the level of the main channel and putting pressure on its levees. The author refers the reader to *The Control of Nature* (McPhee 1990) for details on the series of changes made to the channel.

Rationale for an archaeological perspective

Nothing like the 2005 Katrina disaster in New Orleans has been seen in a city in the USA since the San Francisco earthquake of 1906. Thus, the disaster in New Orleans drew attention from all sides of academia, including social sciences, geosciences, environmental sciences, and, most important, civil engineering. All these fields have concluded that the disaster was typical of the failure to establish a sustainable relationship with the environment. In the context of this conclusion, some archaeologists suggested that the example should be studied to better understand the processes in the past (Bagwell 2009; Dawdy 2006).

Interestingly, although providing many lessons in the broad spectrum of natural and social sciences, the disaster can also teach us lessons in geoarchaeology, particularly in terms of the formation of the record, as well as transformation processes that can be interpreted from the stratigraphy and landforms created by the event.

The case presented here draws on a study (Nelson and Leclair 2006) that by no means had an archaeological or even geoarchaeological purpose, but it provides us with a glimpse of the stratigraphy of the disaster, which provides an interesting view of the combination of natural and cultural features in the sedimentary deposit in the geoarchaeological record in urban contexts, with multiple cases of complexity, equifinality, and multicausality, three of the problems recognized as being typical of geomorphology (Schumm 1991) and geoarchaeology (Chapters 3 and 4).

The "geoarchaeological" context of New Orleans

The Mississippi Delta, and the lower Mississippi valley in general, are highly dynamic environments, where intense changes have occurred in a short time (Aslan, Autin, and

Blum 2005), an aspect that is evident in the geoarchaeological record of the past few millennia in the region (Kidder 1996). This dynamism makes the landscape also extremely young, which is evidence of how the location of New Orleans has changed in the recent past. Only 4000 years ago its location was offshore (Figure 13.5c). The formation of a sand bar (between ca 4000 and 3500 years BP) constituted the first evidence of land in what is today the city (Figure 13.5d–f). This bar, which contained prehistoric occupations (Corbeille 1962) and is known as the Pine Ridge sand bar, has been mapped below the city through coring and excavation (Figure 13.5b).

As the Mississippi Delta prograded into the sea, the sand bar was covered by deltaic and fluvial deposits (Figure 13.5f). Thus, by the time of the city's foundation at the end of the eighteenth century, the area that today the city occupies consisted of a natural level with adjacent low areas of backswamps prone to flooding. The city was founded on the highest and relatively flood-safe surface of the levees, but complications with nature began as the city extended into the backswamps, thus becoming more vulnerable to flooding. The flooding problem was partially solved by upgrading levees, building walls, and pumping out water to prevent inundation. But, as shown in 2005, these measure were not enough to prevent the flooding of the city under an extreme event.

Flood deposits and their geomorphic and cultural contexts

A study by S. Nelson and S. Leclair (2006) revealed some interesting aspects of the flood deposits at one particular location where levees were breached along the London Avenue Canal (Figure 13.6a). A model section from the location in front of house 2 (Figure 13.6b) shows some of the interesting aspects of sedimentation starting with an avalanche carrying a mixture of sediments and fragments of the levees (Unit B), and capped by pulses of sedimentation with variable change in energy (Units C and B) and final deposition of mostly organic, low-density, floating material such as twigs, leaves, and cultural objects (Unit A). Among several interesting results, there are three specific points to be brought to the attention of geoarchaeologists.

First, the formation of the crevasse splays that formed from the levee breaching in the study area differ from crevasse splays in natural environments in two particular aspects. One is that their shape was created in the form of lobes (Figure 13.6a) as opposed to natural crevasses, which have a fan-shaped form. Such a lobate form was created by the layout of streets and houses. Another aspect of particular interest is the planar bedding of deposits (Units B and C in Figure 13.6b), which indicates a relatively low energy involved despite being close to the point of breaching. Most likely the low-energy movement was created as water was slowed down by the buildings. The typical cross-bedded structures and scour structure typical of natural crevasse splays in deposits did occur in certain areas, but they were caused mainly by the influence of certain structures that made the flow cascade or speed up at specific locations (Nelson and Leclair 2006).

Second, the presence of medium sand deposits, with sea sand and sea shells, represents an odd occurrence, because of the distance from the seashore. At this point the channel of the Mississippi does not carry sediments of that sort. Thus, the source of the sand was determined to be the Pine Ridge sand that lies below the channel (Figure 13.5a), which suggests that the force of the storm surge scoured the sand below the channel and added it to the deposits of the splay.

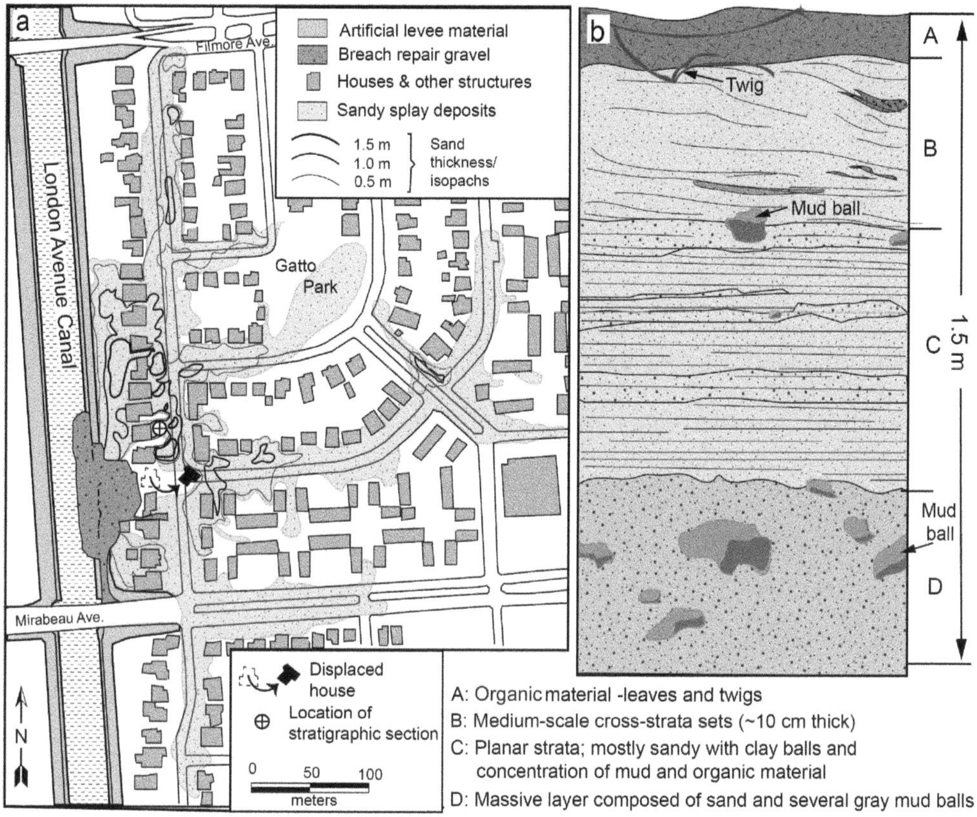

Figure 13.6. a) Splay deposits from levee breach at the London Avenue Canal; b) section showing sedimentary structures in a splay deposit (see locality on the map). Modified from Nelson and Leclair (2006).

Third is the mixture of materials embedded in the sand that included the shells and possibly other objects dating back to that period (see the relative age of the sand bar in Figure 13.5e) mixed with fragments of clay from the levees and modern objects from the local houses. This mixture of fragments datable to different ages is typical of deposits in urban areas, where the force of water may scour and carry materials of different ages. This seems to be a warning regarding the use of artifacts, and even organics, for dating flood deposits.

The lessons for fluvial and urban geoarchaeology

After the disastrous flooding event, little by little the city was cleaned. For geoarchaeology, the cleaning of "mud "and "debris" in essence, meant "erasing the record," which in turn leads to an absence in the record. Therefore, the first lesson to bear in mind is that evidence of disasters in certain settlements, particularly urban environments, can disappear as reconstruction occurs. Had New Orleans been abandoned after the event, the deposits described above would have perhaps allowed future geologists, geomorphologists, and geoarchaeologists to study the disaster through sediments.

As stated by the authors of this study, this flooding is a unique example of the effects of crevasse splay sedimentation in urban areas, which is practically unknown, since most actualistic studies in geomorphology and sedimentology draw on levee breaching cases in non-urban areas, mostly as intentional breaching to restore swamp areas (Nelson and Leclair 2006). But for geoarchaeology this is a unique case of formation processes that could explain many irregularities in environments where natural and cultural processes mix.

Flood deposits in urban archaeology have for a long time created many problems of interpretation (e.g. Addyman et al. 1976; Butzer, Miralles, and Mateu 1983; Brown 1997). Such problems are mainly in reconstructing and dating the events. At the contextual level of setting, the example presented here shows the complexities of one single event, from determining a facies model to using materials for dating. It shows divergence in the record, a problem that, more than geomorphological, is also geoarchaeological.

14

Environmental Crises in the Geoarchaeological Record

INTRODUCTION

As defined in the previous chapter, environmental crises are more complex and often more protracted than natural disasters. Their complexity lies in their causes, which can encompass multiple natural and social phenomena and the multiple events that characterize their development (van der Leeuw 2009). Indeed, the interaction between natural and social phenomena is almost always the factor that defines an environmental crisis, and what differentiates it from other crises, e.g., economic, social, and political. Yet these other crises may play a part in the conundrum of natural and social causes and events involved in an environmental crisis.

The concept of environmental crisis is often associated with events that lead to societal collapse, but such a relationship is a complex one because not all cases of environmental crises lead to collapse. Therefore, it would be important to discuss how environmental crises and societal collapse are related, particularly looking at some of the recent ideas on the subject and some examples.

But the conceptual aspects involved in societal and environmental change, the most important and practical aspect is how environmental crises appear in the geoarchaeological record and how they are interpreted. Although the collapse of the Classic Maya (Case 10.2) is a good example of an imprint left by an environmental crisis in the record, it is important to discuss other aspects of the record at different scales in time and space. For this reason, the two cases in this chapter address issues when environmental crises are clear at one scale but elusive at others (Case 14.1), or when environmental crises are clear in historical records but unclear in the geoarchaeological record (Case 14.2).

ENVIRONMENTAL CRISES AND THEIR RELATION TO SOCIETAL COLLAPSE

Environmental crises and the collapse phenomenon

Many of the examples of societal collapse in the past have occurred as a result of environmental crises, or sometimes as the result of natural disasters that may have triggered environmental crises. However, not all environmental crises lead to societal collapse, and not all cases of societal collapse are the result of environmental crises, but with or without collapse, environmental crises produce a change in society and the environment (Butzer 2012).

Although an environmental crisis has its origin on natural phenomena, such as natural disasters, or climatic shifts, some of the causes that aggravate environmental crises originate within society itself. Such causes include a series of complex social, political, and technological aspects such as institutional failure, corruption, and mismanagement of human ecosystems (Butzer 2012).

As shown in the Yellow River alluvial problems (discussed in Chapter 10), the dynastic institutions failed to solve the problems posed by the environmental crisis, until the problems reached a tipping point and the crisis ended in a series of catastrophes. The ruling elite was unable to address problems related to hydraulic management (Kidder and Liu 2014). Thus, regardless of any catastrophic flood or other natural phenomenon, preparedness through the implementation of proper management is an essential point.

Natural disasters act sometimes as the tipping point that exposes problems of management, either things that were not addressed or not properly addressed. The case of New Orleans (Case 13.2) is a good example of a series of management, technical, environmental, and social problems that were left unaddressed for too long before disaster struck.

Although in modern or historical cases the relations between management and environment are clearer, they are more difficult to define in prehistoric societies, in which cases causality is open to speculation. Sometimes, when the causes of a crisis are not found in the local record they are searched for in the broader geographic context.

Climatic changes, or more properly climatic crises, are invoked often as the cause of collapse. However, the effect of climate on society is rarely direct, often being just part of other environmental and societal issues, that is to say, of an environmental crisis. Many examples in the archaeological record reflect such a relationship between climate and other parts of the environment such as the already-discussed case of the Classic Maya collapse and the case of the Old-World crisis in the third millennium BC (Case 14.1).

The regional environmental crisis that led to the so-called 1177 BCE collapse of the Old World order is a case in which climate played an important role. However, it was only part of a series of other phenomena, including natural disasters and political and economic disruptions, that led to a domino-fashion collapse of states, which, put in a different way, was "the perfect storm of calamities" (Cline 2014: 139).

Looking at the contemporary past

The link between the present and the past becomes more and more important in modern times as we learn more about the current environmental crises through those of the past, and more about the past through the present ones. Thus interest in current human-environmental interactions as a continuum from the past is an idea that has already become part of the archaeololgical and geoarchaeological discourses of the twenty-first century (van der Leeuw, 2009; Kidder and Liu 2014; Butzer and Endfield 2012). Therefore, although the topic of looking at us (the contemporary past) belongs in Chapter 17, it is important to bring up some important aspects of our modern regional and global crises.

In modern times the mechanisms of environmental crises and their effects are different from those in ancient times, in part because of a more globalized world where economies are no longer isolated (O'Sullivan 2008). Communication and interconnectedness between polities, or as K.W. Butzer put it, modern societies (or states), are saved by "information diffusion and socioeconomic integration" (Butzer 2012: 3639).

In the globalized new world, the collapse of a state would impact the macroregional or global economy. Therefore, joint forces sometimes led by international organizations or coalitions of countries intervene at least to try to prevent total collapse of a state or a group of states or prevent the spread across other regions. The case of the Sahel climatic, and certainly environmental, crises of the twentieth century are a good example of aid poured into the region.

Instead of a collapsed state, as in ancient times, a modern state survives despite remaining fragmented, decentralized, and dysfunctional as a state (e.g., Somalia and Afghanistan). Thus, the term *failed state* is the modern equivalent of a collapsed state. Interestingly, many of the failures of these states, although social and political in large proportion, do include environmental crises.

The Syrian crisis that led to the outbreak of the current civil war, which was poorly known outside the country, encompasses several factors, including an environmental crisis allegedly caused by persistent droughts in the decades preceding the conflict (Gleick 2014; Kelley et al. 2015). This narrative suggests that drought in the countryside forced people to move to the cities, where conditions were getting ripe for revolution. Although this explanation seems opportunistic, signs of social distress caused by the drought were already in place in Syria (Juusola 2010) and even in the broader Middle East (Keramat, Meravani, and Samsani 2011).

However, not everyone agrees with the influence of climate as a decisive factor, or even a factor at all in the Syrian conflict (De Châtel 2014; Fröhlich 2016). The argument against the climatic influence lies in the complexity of the environmental, political, and social crisis that has been brewing in the country for decades. This case emphasizes the point raised in this chapter: the complexity of environmental crisis and the processes that lead to collapse.

Interestingly, a different argument to support the indirect influence of climate in triggering the modern Syrian conflict can be built on data provided by another study that shows the inability of the government to address problems caused by the drought (Balanche 2012). This study points precisely to the area of the country that fell under the domination of ISIS. But despite all these little facts, it is difficult to build an argument for conflict and collapse based on climate change – at least at this point.

In the recent past, local crises such as the Dust Bowl of the 1930s did not lead to collapse of society, but only to the collapse of certain systems, in this case of changes in agricultural practices. Also, in this case, however, the crisis represents only a region within a state that has the capabilities to intervene and help remediate the crisis. Although mechanisms such as recovery are different from ancient environmental crises, aspects such as abandonment, migration, land use change, and resilience are similar.

Causality issues in the interpretation of past environmental crises

One reason why societal collapse is such a contentious topic is because scholars often look at common denominators in explanatory models that are designed to fit every case. This same trend exists in trying to characterize environmental crises. The problem with establishing causal links between phenomena of different nature (e.g., seismic, volcanic, climatic, social, economic, technological, and political) lies in the fact that not necessarily one phenomenon has the same repercussions on every society. For example, droughts in the Great Plains of North America and the Sahel had different effects on society. The societal and political context were different, and so were the societal responses to each crisis, despite both occurring in the same century.

Another common aspect discussed as a cause of collapse has been to land degradation, which often involves causal complexities associated with demographics, which is not always the case as numerous examples around the world show (Fisher 2009; Borejsza and Joyce 2017; Brown 2017). Population growth can put pressure on the environment as new marginal or not formerly used areas are open to agriculture as the case of the Classic Maya (Case 10.2), or the case of the plowing of the Great Plains, one of the main causes that precipitated the Dust Bowl crisis (Case 14.2). However, in some cases, it is the opposite; population decline can be detrimental in an environment that has been transformed and dependent on management (Fisher 2009; Brown 2017). Therefore, land degradation as a cause of environmental crisis needs to be carefully assessed case by case.

WHAT DO ENVIRONMENTAL CRISES LOOK LIKE IN THE GEOARCHAEOLOGICAL RECORD?

Unlike natural catastrophes, which leave a clearer mark, environmental crises are more difficult to identify in the record. The evidence left by natural disasters, particularly those of high magnitude (e.g., volcanic eruptions, catastrophic floods, certain storms, and intense droughts with strong wind energy) is perhaps easier to find. In contrast, the complexity of events that constitute an environmental crises make their definition in the record difficult.

Because they are often protracted phenomena, environmental crises evolve. Thus, a crisis may begin as one particular series of events and evolve into others. Take the example of the environmental crisis that led to the Classic Maya collapse: it evolved from a problem of land management to a problem of facing a drought lasting at least three centuries. Moreover, the effects of the crisis affected the different Mayan polities and regions at different times and with different intensity.

It often happens that the effects of an environmental crisis appear differently at different scales. The environmental crisis that led to the collapse of the Near Eastern states at the end of the third millenium BCE (Case 14.1) is an example of a crisis, whose evidence in the record seems clear at some scales but elusive at others. This situation can be a problem when only very few localities or environments across a region are studied. In contrast, the so-called Late Bronze Age Collapse, or the 1177 BCE collapse, for which there is clear evidence at several scales from the levels of site (e.g., destruction and abandonment layers), and landscape (widespread abandonment), and environment (pollen and sedimentation records). Good examples of all these views can be seen locally and in the broad regional context (e.g., Kaniewski et al. 2007; 2010).

The various cases of environmental crises in the past and the present show that the duration and extent of a crisis plays an important role in its definition in the record. Longer crises have more time to leave a mark in high-resolution records, such as depositional systems with low sedimentation rates (e.g., lacustrine sediments), tree-rings, and speleothemes in caves. High-resolution records from multiple lake records and speleothem sequences in the Central Mayan Lowlands region have been of great use in providing an environmental background to the environmental crisis that led to the Maya collapse. In the American southwest, where lake sequences are absent and speleothems very rare, tree rings have been an important source of climatic data to use as a background to some of the crises that have led to the collapse of agricultural settlements in prehistory.

Other records with less resolution but very useful in identifying environmental crisis in the record include alluvial and eolian sequences. But such records may pose problems if not properly dated or not correlated properly across the region, particularly in terms of associating geomorphic events with cultural events, as shown in the example of the Levantine flood plains at the end of the Early Bronze Age (Case 14.1). Care should also be taken on such records because of the time-transgressive nature of their responses to changes entailed by an environmental crisis as the case shown in the effects of the Dust Bowl crisis (Case 14.2).

Finally, it is important to make the point clear that every environmental crisis leaves a particular set of imprints in the record depending on its duration and geographic extent, environment, and socio-political and economic contexts. Although looking at past and present examples of crises helps us to grasp the idea of complexity, it is important to put each case in its own context.

CASE 14.1: THE OLD-WORLD ENVIRONMENTAL CRISIS AT THE END OF THE THIRD MILLENNIUM BC AS SEEN IN THE DEGRADATION OF LEVANTINE FLOOD PLAINS

Background

The environmental crisis of the third millennium BCE constitutes a series of natural and cultural events that led to the collapse of several Early Bronze Age societies across the Eastern Mediterranean and the Near East (the Middle Kingdom and the Akkadian Empire), and South Asia (the Harappan Civilization in the Indus Valley). Because the collapse occurred between 4200 and 4000 years BP, it has been associated with the abrupt global climatic change known as the 4.2 ka event (Weiss 1997; 2000). However, the causes of the collapse and its effects have been debated, with some suggesting that they may be more than just climatic (Butzer 1997; Hassan 1997).

In fact, because of the complexity in archaeological, geoarchaeological, and paleoecological records, rather than a climatic crisis, the events seem to be more associated with an environmental crisis, where climate played a significant role (Cordova 2007). Concentrations of quartz in sediments from the Persian Gulf have been interpreted as the result of dust storms produced by persistent regional dryness (Cullen et al. 2000). However, abandonment and failure of agriculture, and land use changes – even if they were caused by climatic dryness – could have also led to such a crisis (see Case 14.2).

Climatic reconstructions from cave speleothems (Bar-Matthews et al. 1998) do show some decline in moisture around the 4.2 ka event. However, the curve shows drier and more variable events before at several times in the Middle Holocene, thus making the dry phase at end of the third millennium BCE less significant. Some palynological data from the southern Levant show some changes in vegetation at that time, but they are difficult to assess as climatic change because of the transformation of the landscape at the time (Cordova 2007). In the Levant, for example, they show the decline of olive cultivation, which interestingly is replaced by oak (Cordova 2010).

Although some paleoecological records show somehow climatic, if not environmental, change in the late third millennium BCE, little information has been generated from geoarchaeological records. Thus, in this case the discussion will focus on one particular part of the record: the alluvial record of small-stream flood plains in the southern Levant. To place such a regional event in context, the reader is introduced first to the context of the climatic events associated with the 4.2 ka event.

From global to regional records

The effects of the 4.2 ka event are not well known, but they have been identified as a relative cooling associated with an increase in insolation (Weiss 2000). The cooling had an effect on temperatures in the Atlantic Ocean, which in turn affected the development of cyclones and low-pressure systems across the Mediterranean. At the same time, cooling in the Indian Ocean diminished the intensity and influence of the monsoon, affecting the catchments of the Indus and Nile rivers, and impacting their irrigation systems in their lower reaches (Butzer 1976, 1997; Hassan 1997). The Nile originates in an area where rainfall is influenced by the Indian Ocean Monsoon, which, when reduced, makes floods in Egypt insufficient to fill in the floodplain basins and irrigate crops (Figure 10.1).

Under circumstances of reduced precipitation from the Mediterranean, rainfed agriculture in the Upper Mesopotamia and the Levant would be affected, and irrigation systems in the large river systems in lower Mesopotamia would be reduced leading also to salinization (Butzer 1976). Under such a scenario, the climatic influence on agricultural systems is evident.

On the other hand, the global connection is not uniquely linked to the local effects climatic deterioration, but also to a regional economy and trade relations. Thus, failure of agriculture in one polity may affect the others because each polity is dependent on trade with other regional economies, thus collapsing in a domino-effect fashion (Butzer 1997). For example, the Early Bronze Age economies of the Levant were strongly dependent on trade with Egypt, thus economic crisis in Egypt affected them causing their collapse. A similar situation occurred in the crisis of Late Bronze Age Collapse (a.k.a. the 1177 BCE Collapse), which suggests that regional connections were already well established at the end of the third millennium BCE.

Levantine flood plains of medium-order streams: a local response

Even before the widespread idea of a collapse event at the end of the third millennium BCE, evidence of change in small-catchment streams (often third-order streams) of the Levant had been reported in association with the last phases of the Early Bronze Age (Figure 14.1). More research in Transjordan has revealed that widespread stream incision occurred during or after the abandonment of EBA sites (Cordova 2007, 2008).

One particular case is the floodplain incision that occurred in Wadi al-Wala at that time (Figure 14.1). The nearby Early Bronze Age site of Khirbet Iskander has occupations ranging from late Chalcolithic through the Early Bronze Age, phases EBA I, II, III, and IV. During the EBA IV phase, population declined considerably and deserted by the end of the phase and never re-occupied (Cordova 2008; Richard et al. 2010). It is believed that the incision all along Wadi al-Wala and its tributaries may have made floodplain agriculture and irrigation impossible (Cordova 2008). However, given the archaeological records many other sites in the areas east of the Jordan Valley-Dead Sea rift (i.e., Transjordan), remained occupied through the EBA IV, although many of these sites were confined to

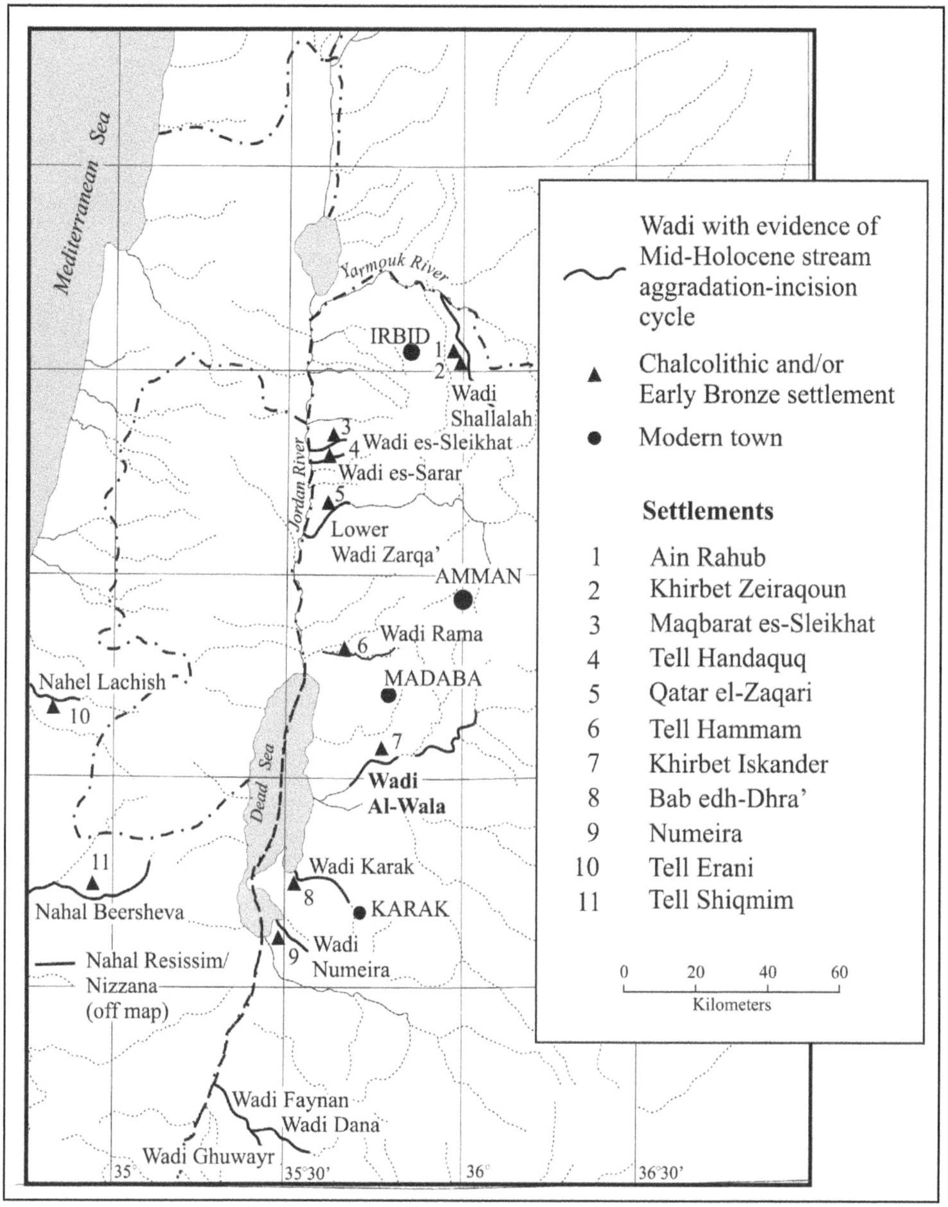

Figure 14.1. Map of the southern Levant with localities mentioned in text (modified from Cordova 2007).

locations near springs and/or streams (Harrison 1997). Still, some of the wadis near these sites show signs of floodplain incision (Mabry 1992; Cordova 2007, 2008). In wadis dissecting the southern plains of Israel, incision events dated to the end of the Early Bronze age were also reported (see Rosen 1995, 2007).

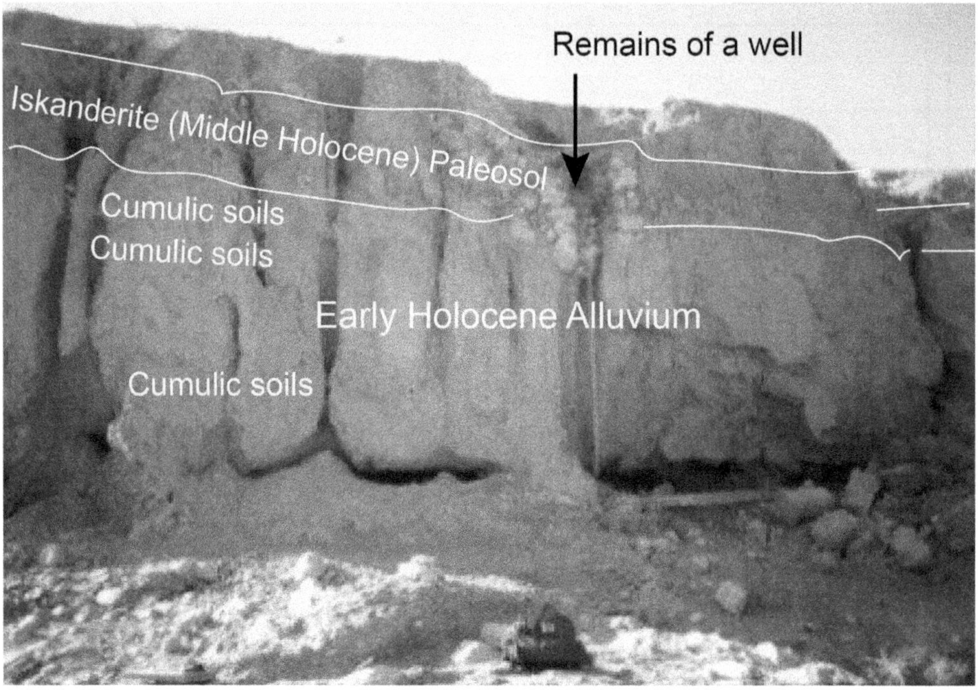

Figure 14.2. View of section in Wadi al-Wala showing the Iskanderite soil and remains of a well. See context of these features in Figure 14.1 (Chapter 4).

The causes of floodplain incision in the cases above may be explained as a function of geomorphic processes in response to changes in the base level. These processes could have occurred only under conditions of lowering of the water table (Figure 14.1). In fact, evidence of wells suggests that before incision the water level was much higher (Figure 14.2).

One explanation for the survival of Khirbet Iskander into the EBA IV phases, unlike others in the area, is that the adjacent flood plain in Wadi al-Wala may have become incised later than in other wadis. A geoarchaeological survey along the wadi showed that a tilted, hard lithologic stratum held the sediments in the form of a natural dam, until it was probably breached, causing the eventual incision of the flood plain (Cordova 2008).

Problems to solve

It is, however, possible that the widespread nature of incision in small streams over a broad area may have actually been triggered by climatic variability and dryness caused by the effects of the 4.2 ka. But the information thus far produced does not provide a 100% certainty that only climatic factors were involved in triggering the process. Land degradation may have also played a factor in triggering the process as it is assumed that when the climatic crisis set in, slopes were already overgrazed and probably with little cover (Cordova 2007). However, this issue is still to be investigated because despite some additional records it is still difficult to determine the conditions of the paleolandscape at the end of the EBIII and EBIV. Pollen data from some of the deposits shows that the spectra is not much different from the present, suggesting that perhaps the landscape was already in a deeply transformed phase, much as it is today (Cordova 2008, 2010).

Perhaps the combination of a climatically triggered effect on an over-exhausted agricultural system in combination with the regional economic crisis contributed to the environmental crisis that drove the last Early Bronze Age communities in the Levant to collapse. But this cause cannot be generalized to all the cases in the region, where records are missing. Besides, even in the studied cases, more detailed dating and micro-stratigraphy would be needed to determine the causation processes of floodplain incision at local scale.

CASE 14.2: THE DUST BOWL IN THE (FUTURE) GEOARCHAEOLOGICAL RECORD OF THE GREAT PLAINS OF NORTH AMERICA

Background

In a relatively short time the Great Plains of North America were transformed from grassland grazed by bison and hunting grounds for several tribes, into cattle grazing range and cropland. This transformation was even faster in the western half of the region, where precipitation is usually variable and for the most part insufficient for rainfed agriculture, namely the area originally covered by the shortgrass prairie (Figure 8.5).

Meteorological records and tree-ring data show that the period between 1900 and 1929 was one of relatively abundant rains, except for a few short droughts (Worster 1979). During this time, encouraged by the US government and developers, farmers moved into the area. The breaking of the protective sod by plowing began to expose soft sediments.

The outbreak of World War I put the USA at the front of grain production and supply, a time that for the plains represented a bonanza, which in turn caused more grassland to be turned into cropland. Even marginal areas of thin soils that were generally unfit for agriculture were plowed, creating the first effects of water and wind erosion (Worster 1979). But as long as rains were abundant the exposed surfaces were covered by crop.

The problem, however, occurred when rains began to fail in the late 1920s, marking the beginning of a 12-year drought extending over a large area, although in fact the regional drought period extends from 1928 to 1940 (Fye, Stahle, and Cook 2003). The unprotected lands, with neither grass nor crops, began to blow, causing spectacular dust storms that rolled over the Great Plains and beyond. Known as the Dirty Thirties, or more commonly the Dust Bowl, this constitutes a well-documented environmental crisis. Taking advantage of the documentation, including climatic data, a study aimed at finding possible evidence of the event in the geoarchaeological record of the future (Cordova and Porter 2015). The strategy, results, and conclusions are summarized below.

The 1930s Dust Bowl in the geoarchaeological record: A research strategy

The aim of this research was to look for evidence of the 1930s Dust Bowl in the sedimentary and geomorphic records in various localities of northwest Oklahoma, southwest Kansas and southeast Colorado, all within the area most affected by the Dust Bowl crisis (Figure 14.3).

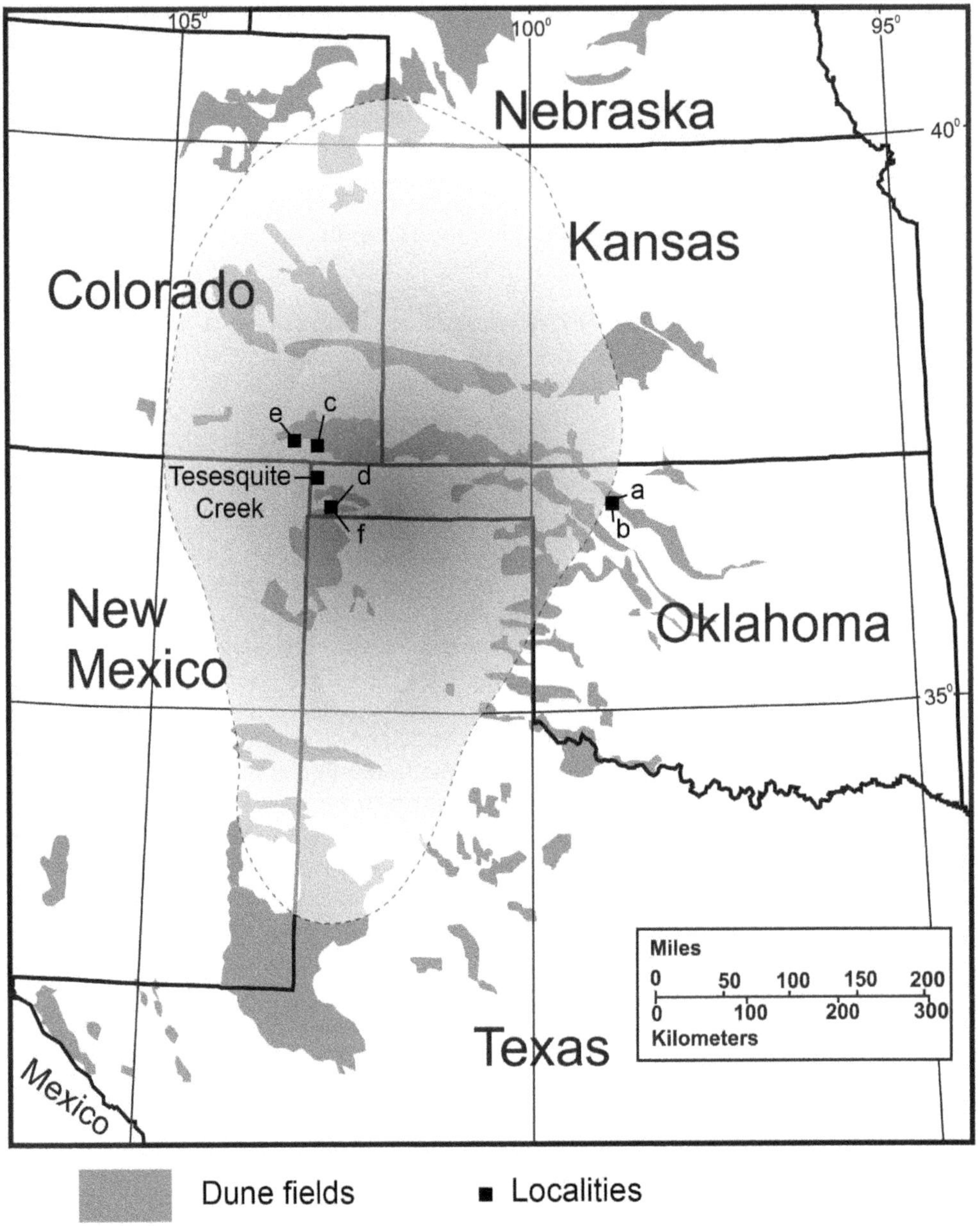

Figure 14.3. Map of the area most affected by the Dust Bowl. Letters refer to the location of sites in the pictures in Figure 14.4. Modified from Cordova and Porter (2015).

The study looked at particular places where the event could be dated, namely areas where objects or constructions pre- or postdating the event could help. Additionally, it used aerial photography and eolian and fluvial records of the area, climatic records, and field observation in areas where local written accounts, old maps, and photographs showed particular changes. Oral and written accounts by local residents were also useful background information particularly at local scale. The main focus of the field research was on artifacts, construction, and landscape features that would help date sediments (Figure 14.4).

Figure 14.4. Artifacts, features, and remains of material culture that could bear archaeological evidence to generations of future archaeologists (Cordova and Porter 2015). a) Fence partially buried by sands active in the 1930s; b) pre-Dust Bowl fence dated using a barbed-wire catalogue; c) abandoned WPA building placed as federal aid in 1937; d) abandoned farmhouse (see history in Cordova and Porter 2015); e) abandoned house with abutting eolian deposit; and f) pre-Dust Bowl machinery around the farm in picture "d."

Archaeological and geoarchaeological evidence of the Dust Bowl event

The study showed that, despite the magnitude of the event, in terms of frequent catastrophic dust storms, the evidence in the geoarchaeological is practically nonexistent. The idea that dust layers would mark the event in the record could not be proved because most of the airborne particles were spread over a large area, including areas as far as the East Coast and the Atlantic Ocean. The only places where dust was preserved was inside abandoned houses, barns, and in the attics of most houses (Cordova and Porter 2015).

Locally, however, evidence does exist in sediments that were not carried very far, mainly sand (Figure 14.5). Thus sand dune mobilization left a record in the form of certain accumulations (small dunes) and blowouts (Cordova et al. 2005; Cordova and Porter 2015). Even so, it is difficult to pinpoint which sand corresponds to the Dust Bowl because the accumulations are insignificant compared with those formed by reactivation periods in the previous two centuries. Furthermore, reactivation of sand in some areas during the 1950s Dust Bowl or other times has also led to reworking of sand dunes.

Only in places where there is datable material such as fences, abandoned farms (sites), and abandoned implements (artifacts) can eolian sands be dated (Figure 14.5). Interestingly, looking at some dune fields on photographs of the late 1930s and early 1940s, many of the mobilized dunes are visible, some of which have become stable in subsequent decades as moisture conditions were stable for the rest of the century (Cordova et al. 2005;

Figure 14.5. Mechanisms of eolian erosion and transport of fine-grained soils and sandy soils (after Pye 1987) during the Dust Bowl event. Modified from Cordova and Porter (2016).

Cordova and Porter 2015). Thus, stabilization of sand dunes by vegetation has been part of the resilient process, at least at a geomorphic level. Nonetheless, the modified dunes are more a local phenomenon, which in the field are difficult to identify as occurring during the Dust Bowl years.

In other cases, the time-transgressive nature of geomorphic systems as in the stream incision created by the 1930s droughts observed in Tesesquite Creek (e.g., Figure 4.5). However, it is possible also possible that other factors may have contributed to incision, as may be grazing pressure concomitant with the rapid agricultural development in the region.

The lessons

Despite being a phenomenon of great magnitude regionally, there is hardly any evidence of the 1930s Dust Bowl in the regional geoarchaeological record. Even locally, the sediments and landforms created by the event are hardly discernible, except at site level, where datable structures of abandoned farms or fences have retained the original blown in sediment. The identification of Dust Bowl deposits and geormorphic features off-site is more difficult, in part because of the time-transgressive nature of most eolian and fluvial processes. In some cases sands may be blending with deposits previous to the Dust Bowl or subsequent, thus forming a palimpsest that would span more than one century.

Provided that the numerous abandoned sites and artifacts are not destroyed or removed, there will be an archaeological record of the environmental crisis, perhaps detectable in archaeological surveys, but more clearly in excavation. In any case, without historical records, it would probably be impossible to define the crisis in the archaeological record. Some may speculate about the social context of the crisis, perhaps invoking the Great Depression, a profound economic crisis that occurred at the same time as the Dust Bowl. Although the two crises had different causes, they were connected and made each other worse.

Thus, instead of being able to identify the Dust Bowl deposits, most likely in the future geoarchaeologists may be able to identify palimpsests of geomorphic features and deposits dated to the "commercial agriculture period" in the Great Plains, thus referring perhaps to the post-1900 economic development of the region.

Like the case of the New Orleans catastrophe, the case of the Dust Bowl shows how complex the combination of natural and cultural phenomena are in events that we usually study in the past. Thus, finding evidence for natural disasters and their effects on society, as well as the development of environmental crisis and the eventual societal collapse, is a tricky business, sometimes leading to misinterpretations and speculations – when the fragmented record is not properly analyzed.

15

Native and Colonial Landscapes

INTRODUCTION

In the late twentieth century the archaeological study of the colonial past began to gain more attention as questions about the legacies in modern society, particularly in Asia, Africa, and the Caribbean, were exposed after the independence process that peaked during the 1950s and 1960s. However, archaeological interests in the colonial past became entrenched in the study of European colonialism, particularly after 1500 CE (Gosden 2004; Orser, Jr. 2014). This conception, however, has been criticized, as cases of colonialism exist in ancient history and prehistory (Rogers 2005).

But even with the extension of the scope across time periods, defining the research extent of archaeology of colonialism is difficult beginning with defining *colonialism* and *colony* (Stein 2005). Alternatively, the archaeology of the colonial encounters fits the idea that most studies focus on the relations between colonizer and colonized, which vary depending on the diverse circumstances of the encounter. Thus, one can find cases of direct settlement by colonizers by displacing or exploiting native populations and their resources as well as cases of indirect rule through the creation of protectorates or areas of influence.

The general problem with the archeology of colonialism and colonial encounters is that there are cases of colonization in almost any ancient migrations. For example, the Bantu migration across central and southern Africa, which would have constituted Bantu colonial enclaves in areas previously inhabited by Khoisan people, is not seen as a colonial encounter. Certainly, one can argue that no central colonial power existed. However, in cases such as the Greek and Phoenician diaspora, which were not necessarily controlled by a central power, the idea of "colonies" is often mentioned. Perhaps the idea of colonial encounters is too Eurocentric.

Therefore, due to the ambiguity of the term colonialism, the terms *colonial encounters* and *cultural encounters* have been proposed (Gosden 2004). Thus, the archaeology of cultural encounters seems to include among other things such encounters that could be broadly defined as colonial encounters or just situations of cultural exchange.

Regardless of the complicated semantics linked to the archaeology of colonial encounters, geoarchaeology plays the same role as in other focus fields of archaeology, that is to say, by applying geoscience methods where necessary to reconstruct formation processes at different scales and reconstruct the ecological context of events. So, why a separate chapter on geoarchaeology and colonial encounters? The answer is that the scope of the human-environmental approach in geoarchaeology encompasses the reconstruction of the ecological context, which, among other things, looks at landscape change. Thus, the explanation of some legacies of colonialism can readily be obtained from the record using geoarchaeological techniques. However, to achieve that objective it is necessary to place the questions of a study in the historical context, which among many could be a colonial encounter.

Two cases are used to illustrate the general topic of the chapter, one typical of Mesoamerica, where a well-established rural system was in place (Case 15.1) and one in southern Africa, where Europeans encountered hunter-gatherers, nomadic pastoralists, and pastoralist-farmers (Case 15.2). The two cases emphasize this idea of change through time as new groups develop new systems, sometimes maintaining aspects of the old order.

HUMAN-ENVIRONMENTAL INTERACTIONS IN THE CONTEXT OF COLONIAL ENCOUNTERS

Regional environmental discourses regarding colonial encounters

The number of cases of colonial encounters is vast throughout history and across the globe, and the many circumstances of the encounters in different climatic zones to the point that talking about the field in general is difficult. This complexity leads to the inevitable fragmentation of the topic into several regional, thematic, and cultural discourses. Thus, one can find topics such as colonialism in Africa, plantation economies in the Caribbean, Phoenician colonies, etc. Unfortunately, environmental change and landscape transformation, the two topics that concern geoarchaeology, may fall within one of the discourses, the regional one being perhaps the most important.

In the Americas, the understanding of the colonial past was more of an understanding of heritage, but in the 1980s and 1990s, this idea took a different turn, as the debunking of the *pristine myth* took place (Denevan 1992). Strong evidence, in many cases originating from geoarchaeological research, challenged the pristine myth. Since then, more information has been produced regarding the long-term transformation of the land in Pre-Columbian America, and although landscape degradation after the European colonization was evident, it was more due to changes in land-use created by planned and unplanned circumstances (Butzer and Butzer 1993, 1995). Paradoxically, most of the degradation in colonial times was due not to land-use intensification, but abandonment created by an increase in mortality among the native population that was infected with European diseases.

Archaeological views of the colonial past in other parts of the world vary, and are more critical of particular situations, depending on the type of colonial development. In those parts of the world where Europeans settled in large numbers, particularly where the native populations were reduced by disease or mistreatment, the discourse is geared more towards understanding land-use changes, either by transplanting land-use models from Europe, or by hybridizing them with the native models (Butzer and Butzer 1995; Butzer 1996). This hybridization was more evident in places where native populations outnumbered European colonists as in many parts of Latin America. In contrast, territories where Europeans settled in larger numbers, decimating and displacing native populations, as in North America, Australia, New Zealand, and parts of South Africa and southern South America, hybrid forms were rare if at all existent. In the latter case the transformation of the landscape was more pronounced.

In other regions, such as southern Asia, Africa, the Caribbean, and Oceania, where Europeans did not settle in large numbers, changes were still evident through transformation of the landscape through exploitation models that combined many local forms or forms imported from other areas at the same latitude. Plantations of non-native crops in the Caribbean with African slave labor is an example of the exploitation of the land where European colonists rarely settled.

ENVIRONMENTAL RESPONSE AND NATIVE AND NON-NATIVE LEGACIES IN THE LANDSCAPE

The number of cases in which geoarchaeology has contributed to understanding change in colonial encounter contexts is vast. Cases of colonization of hunter-gatherer landscapes by agriculturalists, or of colonization of pastoral landscapes by farmers, abound. The European colonization of Australia is a good example of agricultural transformation through farming and pastoralism of a landscape with no previous agriculture.

The ancient Greek colonization of southwestern Crimea (Case 11.1) occurred in a context where an agricultural group already existed. The Tauri practiced some form of horticulture and limited farming, complementing their economy with pastoralism. Interesting, however, is the fact that Greeks introduced a Mediterranean form of agriculture, which did not occupy the same areas farmed by the native Tauri, but areas that had been used more as pastoral grounds (Cordova 2016b).

New forms of "colonial" agriculture in areas where native agriculture existed represent a complex example of transformation, as occurred in many areas of the Americas after Columbus. However, the processes of change vary depending on the area. Where native population was decimated by epidemics, new forms took place, or were amalgamated with the local systems (Butzer and Butzer 1995). In marginal or remote areas or areas with less loss of native population, some of the indigenous systems were slightly modified or remained almost intact (e.g., slash-and-burn agriculture in the humid tropics). For example, the cultivation systems persisting in the Balsas Region of southern Mexico remained almost intact, with only some changes in the crops produced (see Case 16.1, Chapter 16).

The environmental responses to alien forms of land use is a topic of great interest because of the enormous changes that it implied. The arrival of pastoralism in areas where there was no previous pastoral economy has been blamed for many changes, as occurred in most of the Americas. But even in areas where native pastoralism existed, colonialism changed patterns of grazing, thus leading to widespread degradation (Case 15.2).

In the context of colonial encounters, the study of native and non-native landscapes takes many approaches, of which the most common has been to look at geoarchaeological archives in soils and sediments, where evidence of past environmental disturbances can be found. The examination of alluvial landscapes has been a common trend in studying landscape transformation in the Americas, where the destabilization of streams led to depositional and erosional processes, which can be linked to erosional features in the interfluves.

Less common is the trend to look at sedimentary archives in lacustrine deposits, where a series of proxies can be indicative of disturbances caused by land-use changes. Thus, rates of erosion can be assessed through changes in lacustrine sedimentation rates. Changes in water quality (diatoms) and vegetation (pollen, phytoliths, and microscopic charcoal) can also indicate changes in land use.

The number of cases where geoarchaeology has participated in studying colonial encounters at the landscape and environment levels is enormous, particularly if one considers the broader idea of colonial encounters (i.e., pre- and post-dating 1500 CE). Fewer have focused on colonial encounters in antiquity, although some examples can be found particularly in areas conquered by the three big Mediterranean empires (Roman, Byzantine, and Ottoman). As discussed in this chapter's introduction, one problem is that early cases of cultural encounters are rarely seen as colonial encounters.

The two cases shown in this chapter, although involving modern European colonialism, are meant to stress the aspect of landscape transformation through land use change. The Mexican case stresses the idea of land-use change by looking at settlement patterns and their effects on geomorphic processes. The South African case is meant to propose a topic that has hardly been studied from a geoarchaeological point of view, despite containing historical evidence and abundant features in the geoarchaeological record.

CASE 15.1: THE SPANISH COLONIAL LAND SYSTEM ON AN AZTEC LANDSCAPE: AN EXAMPLE FROM THE BASIN OF MEXICO

Background

The key to understanding the processes of landscape change during the transition from late Prehispanic (whether Aztec, Purepecha, or Tlaxcalan) to Spanish colonial rule requires not only searching for evidence in recent changes in soils, sediments, and landform changes, but also searching the archaeological record and early colonial historical sources.

In the Texcoco region of the Basin of Mexico (Figure 11.5), although few sites have been excavated, there is a substantial amount of archaeological survey data. The historical sources include land grant and litigation documents, some of which contain descriptive maps of the landscape. Additionally, some historical accounts by Spanish or Hispanicized native authors provide a broad social picture for the Late Aztec period. Using a methodological framework combining field geomorphological and soil data with archaeological and historical data, the reconstruction of landscape transformation processes in Aztec and Spanish colonial produces information regarding the effects of land use changes on the landscape (Cordova 1997; Cordova and Parsons 1997; Cordova 2017).

Research and results

In general terms five phases of rapid sedimentation have been identified in the valleys. The last three are dated to the Late Holocene. Unit C is dated to the Preclassic period and is capped by a well-developed soil with apparent distribution throughout the Basin. This soil has phases that are cumulic, but the top horizon is well developed and contains Preclassic and Classic materials (roughly 800 BCE to 800 CE. Unit D is an alluvial unit dated to the Epi-Classic and early Postclassic period, roughly between 800 and 1200 AD. This unit is capped by another soil horizon that contains fragments of different Postclassic ceramics, but predominantly Late Aztec ceramics (i.e., Aztec III-IV). Unit E contains a mixture of light alluvium, showing different facies, from overbank to crevasse splays, and is capped by the modern soil surface.

Of the three alluvial units, Unit E is the thickest and the most widespread, suggesting a massive or relatively short event (Unit E is Unit 1 in Figure 3.4b). Some streams in the upper reaches of the alluvial plains of Texcoco incised after the event, while others in the plain showed evidence of avulsion. Extensive written documentation on flood events in the region, some of which were catastrophic, exist for the late 1500s and early 1600. Constructions of projects for flood prevention also exist, as is the case of Acolman Dam, north of the Texcoco region, in the early 1600s (Cordova 1997). Another short alluvial crisis with sedimentations and avulsion cases occurred in some areas in the 1770s and 1780s, but it was more local and, given the stratigraphic positions, such deposits have been arbitrarily merged with Unit E (Cordova 2017).

The massive sedimentary packages of Unit E were caused by the extensive removal by erosion of sediments and soils from areas upstream (Figure 3.4a). Therefore, the causes of the phenomenon had to be looked for in the piedmont area. This suggested the revisiting of the original survey maps and notes and a series of colonial maps and documents (Cordova 1997; Cordova and Parsons 1997).

Complex land-use pattern changes as a cause

When conquistador Hernan Cortes related his arrival in the Basin of Mexico to the Spanish emperor Charles V, he described the settlement of the Texcoco Region as being disperse and without any city (see full citation in Cordova and Parsons 1997). This settlement pattern is reflected in the distribution of Late Aztec materials on the piedmont of the Texcoco region in the archaeological survey descriptions and maps. Sites have been mapped as extensive areas of domestic units and ceramics dispersed on the fields delimited only by gullies and areas of erosion (Figure 3.3).

Many of these areas were apparently rural, but according to some sources they were tied to certain polities that in records appear as cities – or the way the Nahuatl word *altepetl* is often translated. However, the idea of city in the sense of agglomeration in this part of the basin is different from the European idea of city. What today is represented by towns (e.g., Texcoco, Huexotla, Coatlinchan, and Tepetlaoxtoc) seem to have been centers where the elite lived and where the civic and religious structures were located. The rest of the population, constituting mainly the class of *macehualtin* (commoners), lived dispersed in the landscape.

However, the urban centers we see today in the Texcoco region with the same Aztec names are the result of a settlement policy implemented in the late 1500s. Known as *congregaciones*, this policy meant to concentrate the dispersed population into towns. The population had by that time been reduced considerably due to the epidemics. Thus, the Spanish authorities attempted to concentrate the population into towns for taxation and religious purposes. The result was a strong nucleation of the population. Although nucleation did exist at some periods in Prehispanic times (Sanders, Parsons, and Santley 1978), the so-called nucleated settlements were far less concentrated than the nucleation in the Spanish concept of settlement (Cordova 1997).

The implications of the depopulation of the countryside by the epidemics throughout the 1500s and by the policy of *congregaciones* at the end of the century left the landscape unoccupied, a change that was detrimental to the landscape because years of Aztec rule had reclaimed this area by creating soils on barren volcanic tuff (*tepetate*) and retaining sediments in terraces and check dams. Once abandoned, these structures began to deteriorate and break open, letting sediments loose to be removed as the process itself also created cascading streams, whose force eroded more material, carrying it downstream. The problem became accentuated when, after long periods of drought, copious rains fell on the region, a situation that characterized the weather of first three decades of the 1600s when abundant rains caused flooding and fast sediment accumulation downstream (Cordova 1997).

The problem was exacerbated by the sudden rise of the lake levels, which made the sediments back up onto their flood plains, thus creating the massive amounts of fine-grained overbank sediments that characterize most of Unit E described above. This process was exacerbated in areas of convex flood plains where water in the channels run slightly above than areas beyond the levees (see model in Figure 3.5). Thus, settlements and fields in the low areas of the plains were inundated and eventually covered with sediment (Cordova 2017).

It is not clear how much the introduction of livestock in the area may have contributed to making areas upstream vulnerable to erosion, but it is possible that goats and sheep could have impeded the regeneration of vegetation on the abandoned areas. Clearly, land grant documents (*mercedes*) show that some of the abandoned lands became grazing grounds, but there is no data on how intense grazing was. However, the sole abandonment of structures would have been enough to create the massive transport of sediments downstream under conditions of long drought followed by heavy rains.

Conclusion

The landscape change in the Texcocan piedmont entailed a rapid change from dispersed to nucleated settlement under conditions of variable change and livestock grazing in an area highly vulnerable because of the underlying hard pan. Cases of nucleation and dispersion existed in previous times (Figure 15.1), and it is possible that changes occurred in this area, as nucleation in some parts of the basin may have left others abandoned. Maybe the abandonment that occurred during the Late Classic Epiclassic could have led to the erosional phase that produced the sediments of Unit D in the Texcocan flood plains (Cordova 1997, 2017).

The geoarchaeological record shows that Late Aztec farmers in fact reclaimed the already destroyed land (Cordova 1997; Cordova and Parsons 1997). Dispersion was perhaps the only way to reclaim the land and maintain the structures that retained the sediment (Cordova and Parsons 1997). The pattern seems to be recurrent through the mountainous regions of Mexico, particularly in areas where hard sediments underlie soils as in Tlaxcala and Oaxaca (Joyce and Borejsza 2017).

Figure 15.3. (a) Badland landscape on a high terrace of the Sundays River near Graaff Reinet (see location on Figure 15.2); (b) alluvial deposits on a lower terrace of the Sundays River. Unit 3 has several paleosols and cumulic soils with hearths; units 2 and 3 are two relatively fast depositional events. Photographs by the author.

pastoralism and land use changes of the past three centuries, no study has attempted to date erosional features and alluvial fills to provide evidence support to the process of land degradation, or landscape transformation in general.

One that has remained elusive in the historical ecology of South Africa is the potential impact of native pastoralism on the landscape. Pastoral nomads arrived in the Cape Region in about 2000 BP, according to some dates from various sites (Bousman 1998). That is about 1700 years before the Dutch began moving into the interior of the continent. Although records from lakes in the Cape attest to ecological changes created by changes in land use under European rule (e.g., Scott and Bousman 1990; Baxter and Meadows 1994; Neumann, Scott, and Bamford 2011), the effect of land-use changes by the arrival of native pastoralism remains elusive in the pollen records of the Cape Region, and in general of most of the country. This situation certainly contrasts with the records of Lake Ngami Botswana (Case 9.2) where the impact is clear. Thus, the first issue is to study the relationship between pre-European pastoralists and their landscape, an issue that has to be linked to a much broader question about native and non-native land-use changes (see next section).

The problem: Defining the problem of landscape transformation in colonial South Africa

Hypothetically, one can think that the Khoi pastoralists in the cape practised a more sustainable form of pastoralism, or perhaps a low-intensity pastoral activity that would not leave a mark in the paleoenvironmental record. Alternatively, there has not been a proper focus on that particular event, the arrival and adaptation of Khoi pastoralists in the region, in the records. This obviously means focusing on high-resolution records for that period, but also a series of sources in the paleoenvironmental record – including the geoarchaeological record.

Perspective can also change the way native changes on the landscape left an impact; the European impact was so strong that it belittles any changes left by earlier impacts, pastoral or hunter-gatherers, in the record. Therefore, multiple hypothesis can be posed to explain the possible effects of native pastoralism.

However, not all cases of erosion in these regions could be attributed to the management of the land by European colonists. Many lands in these regions are highly susceptible to variable climate (Meadows and Hoffman 2002). Thus, a more comprehensive study looking at human changes and climate variability should address the general issue of impacts of land-use changes, whether native or European.

Badly eroded areas exist also in in other parts of South Africa, such as the Highveld and the lands south of the Drakensberg Mountains, namely the Ciskei, Transkei, and Zululand. However, it is believed that although this area had farming and pastoral economies, most of the erosion (some in the form of gullies and badlands) is more recent, and linked to a form of colonialism different from that seen in the western parts of the country. Recent erosion in native lands is evidently the result of the division between commercial and non-commercial land and segregation, which in turn made the native lands have higher population densities that in turn were detrimental to soils (Meadows and Hoffman 2003).

Forced to graze in small areas since the times of British administration, native pastoralists caused the damage that characterizes areas, particularly native areas. These also include homelands or bantustans, the name for territories where the Apartheid government concentrated large amounts of the African population. Even today, after more than two decades of the abolition of the bantustans, the borders can be still seen as the former native homelands are characterized by rills and erosion. The ancestors of pastoralists in such areas practised long-range pastoralism, which in the modern era became confined to small areas, those putting pressure on vegetation and soils, particularly in highly vulnerable areas (Meadows and Hoffman 2003).

A proposal: Research strategy

The problem of linking erosion features, sedimentation, and stream incision with changes in land use is a complex one, particularly where the record is dispersed and scarce. This is the case of South Africa, where despite historical records attesting to some landscape changes, there are practically no field records to prove the causality of early colonial land degradation.

Historical records and, where available, high-resolution climatic and paleoecological records, provide the background for a study of erosional features and alluvial deposits that could help determine the causality processes of landscape change beginning with the arrival of native and European pastoralism. Such as study could focus on key geomorphological localities like the one in Figure 15.3. One important aspect in the records is to target those that are likely to provide good resolution for at least the past 4000 or 3000 years.

16

Geoarchaeology and Modern Traditional Societies

INTRODUCTION

The study of human-environmental relations of the past has inevitably drawn references to groups broadly defined as modern traditional societies. In archaeology, the practice of studying modern traditional groups, ethnoarchaeology, originated under the broad umbrella of the New Archaeology movement, which inevitably influenced geoarchaeology, thus leading to the field known as *ethnogeoarchaeology*, the main subject of this chapter.

Before discussing the methodological aspects ethnogeoarchaeological research, it is important to briefly define the concept of modern traditional societies within the framework of the study of human ecological context. Also important as an introduction to ethnogeoarchaeology is to place it in the context of other empirical and actualistic approaches, i.e., experimental geoarchaeology and the geoarchaeology of the contemporary past. It is also important to discuss the areas where ethnogeoarchaeology has succeeded and areas where more research is needed, and the limitations of its practice in the interpretation of the geoarchaeological record.

Because most research studies within the ethnogeoarchaeological approach focus on formation processes at the contextual level of site, it is necessary to explore the potentials at higher levels. Thus, the case presented in this chapter is meant to show the potential that the approach has at the level of landscape. The case study refers to the seasonal construction and destruction of traditional irrigation installations in the Nahua region of the Middle Balsas Basin in southern Mexico (Case 16.1).

DEFINITION OF CONCEPTS AND RESEARCH FIELDS

Modern traditional societies

The term *traditional society* has been broadly used in anthropology and sociology to differentiate certain cultural groups from modern industrial groups. But since the 1970s, the term has been criticized on the basis of its simplicity and unilinearity by pointing to the traditional vs modern dichotomy, often expressed as traditional vs industrial or traditional vs urban (Eisenstadt 1973). It is true that societies that are called *traditional* are no longer totally isolated from the industrial world, let alone from the globalized world. Thus, the definition has evolved to define traditional societies in terms of their attachment to their past, customs, and habits (Langlois 2001).

The matter of tradition and traditional societies in terms of human-environmental relations has been addressed widely by anthropologists, archaeologists, cultural ecologists, and other areas under the umbrella of the broader ecological paradigm in the social sciences. Environmental anthropologists refer to "traditional knowledge," particularly in terms of the use of resources and the environment (Townsend 2000). Still, the aspect of tradition is debated particularly when considering relations with the modern world, an aspect that does matter to archaeologists, who use the traditional knowledge as the basis for the research field of ethnoarchaeology (see below).

For geoarchaeology, whose view of traditional knowledge is placed within the frame of geosciences, ecology, and environmental sciences, the "traditional" part is much less contentious than for archaeology and anthropology. Still, because of problems with interference of the non-traditional aspects of the modern world, results obtained from research in traditional cultural contexts, i.e., the case of ethnogeoarchaeology, should be used purely as reference to interpret processes in the record.

In view of the discussion above, the term used here is *modern traditional society*, in which the term "modern" specifies that a social group is still attached to customs and traditions of the past in a modern context. Thus, modern traditional societies include a variety of groups that for the most part base their subsistence on non-mechanized agriculture, pastoralism, and/or hunting and gathering.

Indigenous groups in the Americas may be cited as examples of modern traditional societies based on language, culture, and even physical characteristics, all of which are attributes that not necessarily make them traditional societies. Take the case, for example, of Mexico, where for statistical purposes an indigenous person is defined as an individual who speaks a native language at home. That is, if one speaks, for example, Mixtec at home, then he or she is considered indigenous. But someone who has 100% Mixtec blood but does not speak the language is then considered mainstream Mexican or Mestizo. In the USA, on the other hand, the definition is better defined on ancestry, which in some cases becomes really complicated as people with various percentages of Native American blood, so to speak, often claim legal rights, fellowships, or simply recognition.

Despite the different conceptual differences based on ethnicity and ancestry, the view of a modern traditional society refers to aspects of their relation with the environment, which is the matter that concerns the study of ethnogeoarchaeology. In some cases, modern traditional societies may subsist on non-traditional activities, but still engage in some form of subsistence practices such as gardening or hunting for supplementary income or nutrition, or in extreme cases for ritual or even nostalgic reasons. Nonetheless, a particular subsistence practice may be abstracted to explain formation processes in the record.

The rise of ethnoarchaeology

Although ethnographic data have been used by archaeologists in the interpretation of the past for some time, it was not until the early 1960s that ethnoarchaeology was consolidated a methodological approach within archaeology (David and Kramer 2001; Watson 2009).

Essentially, ethnoarchaeology encompasses the collection and use ethnographic data to interpret archaeological societies focusing on material culture, the use of space, and resources (David and Kramer 2001).

The increase of ethnographic studies in archaeology was not welcomed by ethnologists, who saw the ethnographic work of archaeologists as mere ethnographic analogies. But it is the distinction between ethnographic study and ethnographic analogy that makes the difference between ethnography and ethnoarchaeology (Watson 2009). "Ethnographic analogies – whether simple or complex – are trial formulations (hypotheses, models) subject to testing like all such propositions" (Watson 2009: 5).

However, the analogy issue has received substantial criticism also within archaeology itself, particularly from the post-processualists. One such criticism has been referred to as "the tyranny of the present," or the idea that modern traditional groups are now different from those of the past essentially because of the modern world surrounding them (Wobst 1978; Shennan 2004). But despite the criticisms, it seems that ethnographic studies are still a strong source of reference for the interpretation of past societies.

Ethnoarchaeology and the ecological context

Ethnoarchaeological methodologies and theories in archaeology and the studies that they produced have had a tremendous impact on environmental archaeology (Shackley and Reitz 2012). In part this is because ethnoarchaeological data has provided substantial information on resource procurement and, in general, information on collective behavioral patterns with regard to environmental factors, certainly alluding to the ecological context. However, ethnoarchaeology fails to provide many aspects of a spatio-temporal nature, particularly because it focuses mainly on behavioral aspects at the level of occupation (i.e., site), often feeding into aspects of site formation, and less on scales involving landscape evolution and environment.

Additionally, many aspects particularly linked to geosciences (mineral world, landforms, weather, climate, etc.) and aspects that involve longer periods of time (i.e., *longue durée*) and links to more regional and global scales are aspects that do not fall in the scope of ethnoarchaeology, as its focus is mainly on material culture, which is often studied at local scales.

Geology and other geosciences offer the technical tools to solve problems such as lithic and mineral sourcing, chemical residues, and site taphonomy. Additionally, geography, geomorphology, and soil science offered many methods and techniques to study the landscape and the environment, including processes of long duration. All of these contributions are essential for studying the ecological context of the modern traditional societies studied by ethnoarchaeology. Consequently, these interdisciplinary developments converge in the creation of a field that would combine methods and techniques of geoarchaeology within the study of modern traditional societies, that is, *ethnogeoarchaeology*.

ETHNOGEOARCHAEOLOGY: DEFINITION AND SCOPE

Ethnogeoarchaeology, or, less commonly, geoethnoarchaeology, might be broadly defined as the geoarchaeological methods to answering questions posed by ethnoarchaeology.

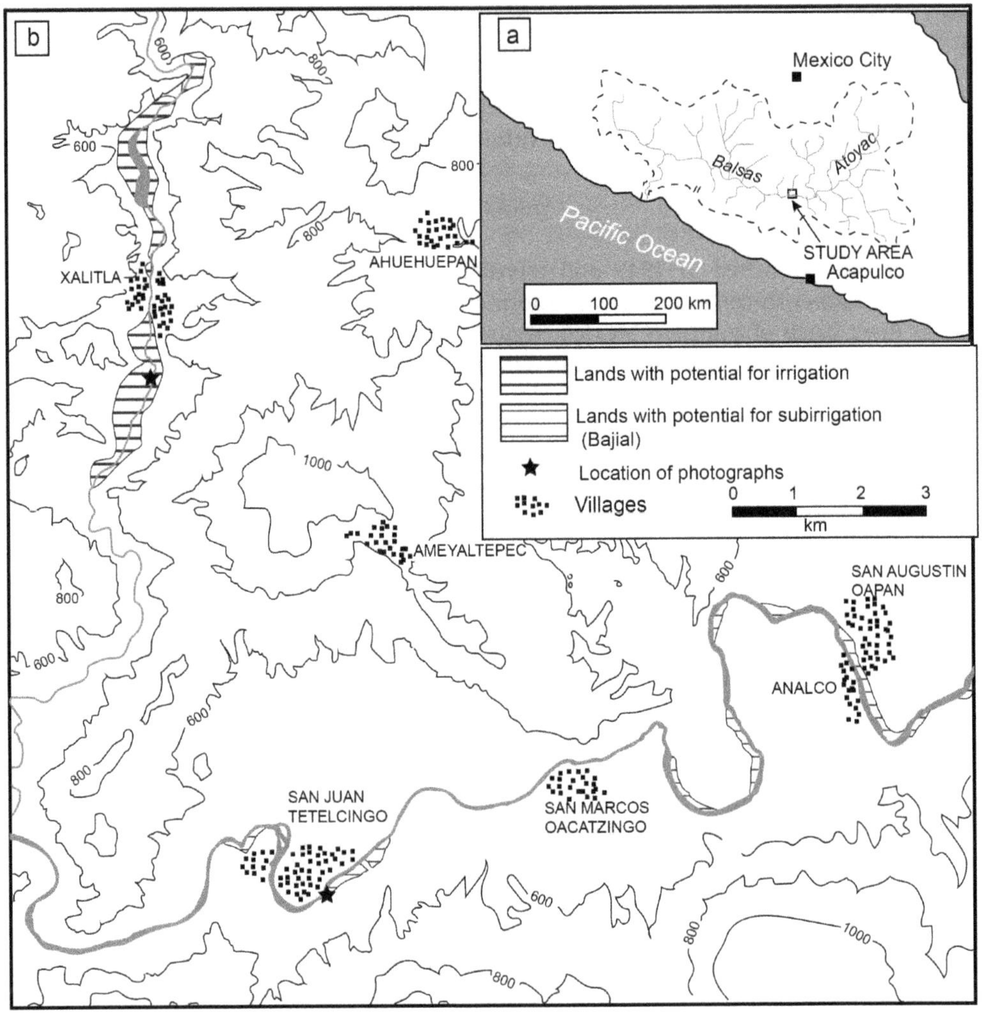

Figure 16.1. a) The Balsas Basin with the location of the study area; b) the study area with the two main streams studied: Tepecuacuilco and Balsas (at this location also called the Mezcala River).

The area has a long history of settlement that predates the Nahua, as testified by large sites dated to the Preclassic, some of which have significant Olmec influences, and to the Classic, some with civic structures and ball game courts (Schmidt and Litvak 1986). Evidence of an excavated canal, allegedly one of the oldest in Mesoamerica, was found in association with Preclassic archaeological material, suggesting the existence of irrigation, probably from a dam built nearby (see comments and discussion in Doolittle 1990). However, the sole finding seems disconnected or out of context, and does not permit reconstruction of an irrigation system.

In an area where precipitation is considerably reduced by the rain-shadow effect of mountains surrounding the basin, the development of sites during the Preclassic or Classic would have necessitated some form of irrigation. But other than the canal mentioned above, no visible evidence of irrigation predating the late Postclassic has been found.

Economy and irrigation in the studied valleys

The Middle Balsas (i.e., the Mezcala) is located in the State of Guerrero, not far from the old Mexico-Acapulco highway (Figure 16.1b). A semiarid region with less than 400 mm of rain a year and steep slopes, the area is barely fit for agriculture. Rains fall within a short period (July–September), during which fallow agriculture takes place in the highest parts of the mountains where the tropical deciduous forest is slashed and burned. Other forms of short fallow agriculture are possible in very limited areas of old river terraces and paleo-volcanic and planar surfaces (Cordova and Vázquez 1991).

Canal irrigation and subirrigation takes place only during the dry season, because it uses areas of the floodplains and channel that are often flooded during the rainy season (Figure 16.2). However, each irrigation system is developed at slightly different times depending on the dry-season regime of the streams. The canal irrigation system along the Tepecuacuilco River is created by building weirs across the main channel and digging canals to irrigate fields that are not possible to farm during the rainy season because they are exposed to frequent flash floods (Figure 16.2a; Figure 16.3). The process weir and canal construction begins in November and crop is planted before the end of the year. Harvest is usually done in April before the rivers rise again as rains begin earlier in the highlands. The main crops are maize and beans.

The subirrigation system on the Balsas River bed, or *bajial*, begins much later, and follows the drop of the water level in the channel (Figure 16.2b). It uses the enriched silts left by the seasonal flood of the streambed and the moisture (Figure 16.4). It is often used to irrigate crops such as flowers, chili peppers, cantaloupes, and water melons. As the underground moisture drops, crops in their last phases are irrigated by carrying water from the river in buckets.

The important aspect to recognize here is that during the rainy season, all the work done through investment in the canal irrigation system is lost as flash floods, characteristic of the season, destroy weirs, canals, and fields. The system has to be rebuilt again. The same occurs with the less-investment subirrigation system.

Today, these irrigation systems in the Balsas Basin are disappearing. In particular, for this area, subsistence is being replaced by cash coming from the successful handicraft business (Cordova and Vasquez 1991). Irrigation today is only practised by a very small part of the community and now it is in very small areas, and mainly for crops that have more to do with traditional rituals, particularly flowers and chili peppers. Other systems such as slash-and-burn and fallow agriculture produce additional crops for subsistence and local consumption. However, the documents studied by Armillas (1984) show that the systems were important forms of subsistence along with fallow agriculture in the sixteenth century, suggesting that at one point the systems were an important source of crops for subsistence, tribute, and trade.

Geoarchaeological significance

Because the structures for the canal irrigation and subirrigation systems are built every year and destroyed by the floods in the rainy season, they are absent in the record. Thus, the first

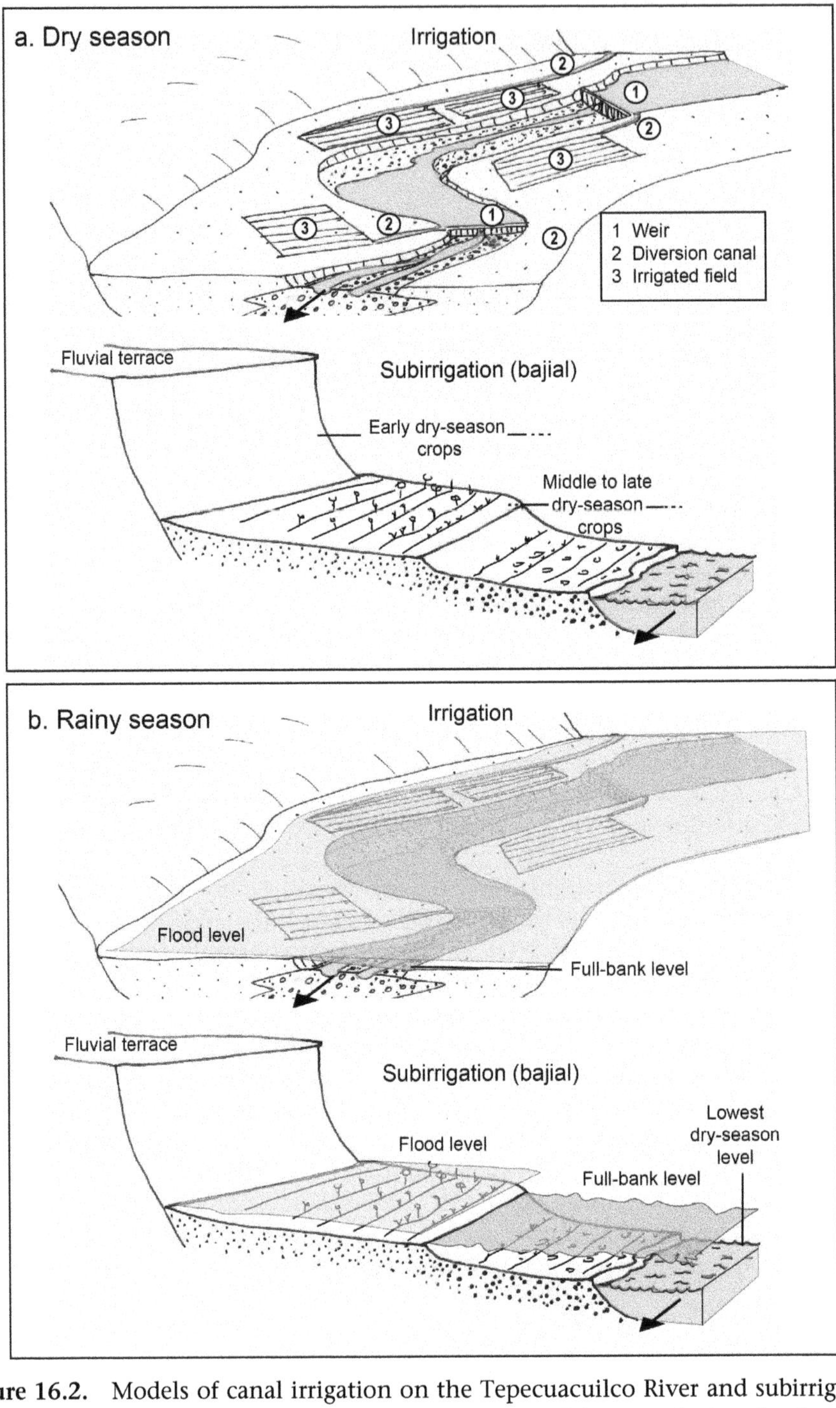

Figure 16.2. Models of canal irrigation on the Tepecuacuilco River and subirrigation (*bajial*), on the banks of the Balsas River. (a) Processes and features during the dry season; (b) processes during the rainy season.

Figure 16.3. Photographs of weirs and canals on the Tepecuacuilco River. Photographs by the author.

17

Geoarchaeology of the Contemporary Past

INTRODUCTION

Geoarchaeology, with its strong focus on landscape and environment, is slowly finding a path towards the objective of reconstructing the recent past and the present, an idea conceptualized in the so-called archaeology of the contemporary past. This trend has gained tremendous momentum since the turn of the century, thus constituting a field with numerous applications to social and environmental problems.

Beyond their applicability, geoarchaeological studies of the contemporary past have an intellectual value. On the one hand, they provide an insight into how we (our current society) will be seen in the future. On the other, like its actualistic counterparts – experimental geoarchaeology and ethnogeoarchaeology – the geoarchaeology of the contemporary past helps clarify processes and complexities in the formation of the record (Table 16.1). Cases such as the 2005 flood deposits in New Orleans (Case 13.2), the sedimentary record of the Dust Bowl (Case 14.2), and the information produced by paleotempestology (Chapter 13), are examples of how the geoarchaeological record is formed, and how its parts become invisible and absent.

Given the examples other chapters, the present chapter includes only one case that illustrates the potential significance of the archaeology and geoarchaeology of the contemporary past. The case presents some aspects of the environmental legacy of the Soviet period in Crimea and its preservation in the archaeological record of the future (Case 17.1). But before discussing this example, this chapter attempts to define the contemporary past as seen by its proponents in archaeology and in terms of its potential in geoarchaeology.

THE CONTEMPORARY PAST DEFINED

Although archaeology is a science that studies the past, its theories and methods are applied to reconstructing a variety of aspects of modern life using archaeological methods (Rathje 1989; Buchli and Lucas 2001). This reconstruction includes the study of modern garbage dumps, war crime mass graves, and abandoned industrial complexes, to name just a few (Harrison and Schofield 2010). The reasons why we look at the present from an archaeological point of view seem too intellectual. However, as detailed by the proponents of this emerging field in archaeology, the study of the contemporary past provides a deeper insight into modernity.